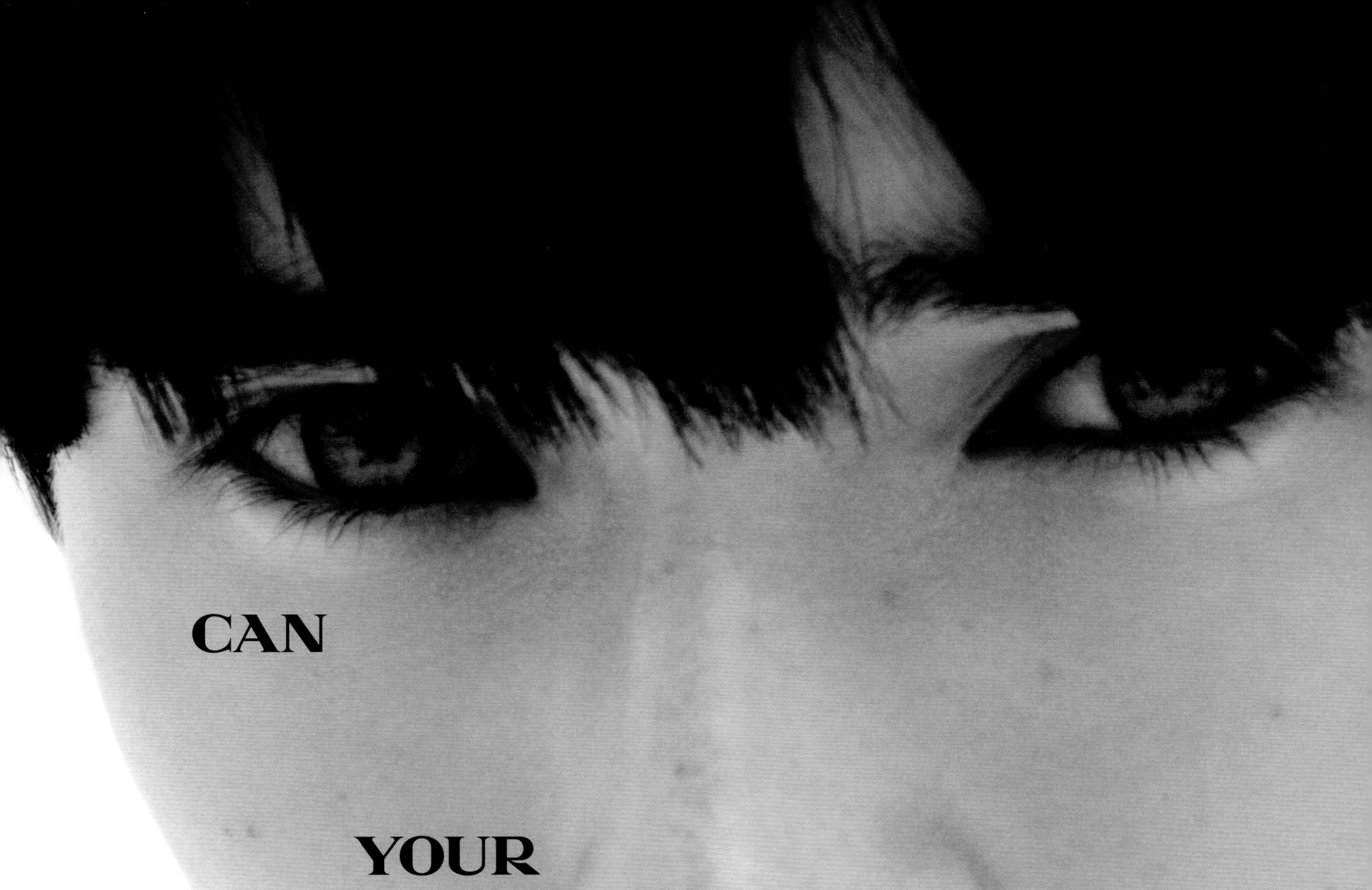

CAN

YOUR

EMOTIONS

BE

ENGINEERED?

ANSWER
AT
PRADA
.COM

SAINT LAURENT
GAUGE
SPRING SUMMER 21 COLLECTION
YSL.COM

**EDITOR IN CHIEF & CREATIVE DIRECTOR** KAORU ŚASAKI

**GENERAL SENIOR EDITOR** KAORI ODA

**FASHION FEATURE EDITOR & PROGRESSIVE MANAGER** KAORI USHIJIMA

**ASSISTANT EDITOR** HIRAME MIYAHARA

**CO-FASHION EDITOR** SHINO ITOI

**CONTRIBUTING FASHION EDITORS** RENREN, NAO KOYABU

**CONTRIBUTING FASHION FEATURE EDITOR** TATSUYA MIURA

**TRANSLATOR** REIKO CRYSTAL LOUCKS

**ART DIRECTOR** KAORU ŚASAKI

**ASSISTANT GRAPHIC DESIGNER** HWANHEE JEON

**GRAPHIC DESIGNER** TAKUHIKO HAYAMI @conte.

**PRINTING DIRECTOR** NOBUMASA KUDO @JAPAN STANDARDS
**PRINTING PROGRESSIVE MANAGER** NORIYUKI IWAMOTO @SOHOKKAI CO., LTD.

**ADVERTISING MANAGER** MINAKO MEGURO minakomeguro@commons-sense.net

**CONTRIBUTORS**
AKIKO SAKAMOTO, BABYMARY, CHIKASHI SUZUKI, COOKIE!, DAIKI KANECHIKA, GO UTSUGI,
ISSEI MAMEHARA, JUNKI KONO, KAGIRU, KAI DE TORRES, KAIGA NAKAMURA, KATSUMI KAWAHARA,
KAZANE, KEIGO SATO, KENICHI YOSHIDA, KENJI YAMAUCHI, KIO DE TORRES, KOHEI, KUNIO KOHZAKI,
MASANORI USHIKI, NAOHIRO TSUKADA, NICO, REN KAWASHIRI, RIHITO ITAGAKI, RINTARO, ROY,
RUKI SHIROIWA, SHIN OKISHIMA, SHINICHI MIKAWA, SHION TSURUBO, SHO YONASHIRO, SHOSEI OHIRA,
SHUCO,SHUHEI UESUGI, SUKAI KINJO, SYOYA KIMATA, TAKUMI KAWANISHI, TAKUYA UCHIYAMA,
TAWARA, THE MISTRESS 5, TOKYO DANDY, YUJI WATANABE, YUME IPPEI, YUTO KUBOTA

**SPECIAL THANKS** GLACE LIBERTY ONPAMIDO

**PUBLISHER** KAORU ŚASAKI

**ISSUE DATE** 27th April 2021

**PUBLISHING** CUBE INC.
4F IL PALAZZINO OMOTESANDO 5-1-6 JINGUMAE SHIBUYA-KU TOKYO 150-0001 JAPAN
tel. 81 3 5468 1871 fax. 81 3 5468 1872
info@commons-sense.net

**DOMESTIC DISTRIBUTION** KAWADE SHOBO SHINSHA
2-32-2 SENDAGAYA SHIBUYA-KU TOKYO 151-0051 JAPAN
tel. 81 3 3404 1201 fax. 81 3 3404 6386
www.kawade.co.jp

**INTERNATIONAL DISTRIBUTION** NEW EXPORT PRESS
www.exportpress.com

**PRINTING** SOHOKKAI CO., LTD.
<TOKYO BRANCH>
2F KATO BUILDING 4-25-10 KOUTOUBASHI SUMIDA-KU TOKYO 130-0022 JAPAN
tel. 81 3 5625 7321 fax. 81 3 5625 7323
<HEAD FACTORY>
2-1 KOUGYOUDANCHI ASAHIKAWA HOKKAIDO 078-8272 JAPAN
tel. 81 166 36 5556 fax. 81 166 36 5657

Copyright reserved ISBN978-4-309-92224-9 C0070

**www.commons-sense.net**

**commons&sense man ISSUE31** will be out on 27th September 2021

ISSUE30 CONTENTS

## ANYONE WHO HAS NEVER MADE A MISTAKE HAS NEVER TRIED ANYTHING NEW

## OUR RICHES, BEING IN OUR BRAINS.

## INTELLECTUALS SOLVE PROBLEMS, GENIUSES PREVENT THEM.

## LIFE IS A TRAGEDY WHEN SEEN IN CLOSE-UP, BUT A COMEDY IN LONG-SHOT.

## LEARN FROM YESTERDAY, LIVE FOR TODAY, HOPE FOR TOMORROW. THE IMPORTANT THING IS NOT TO STOP QUESTIONING.

## MARCH 11TH

## THE VALUE OF A MAN SHOULD BE SEEN IN WHAT HE GIVES AND NOT IN WHAT HE IS ABLE TO RECEIVE - INTERVIEW WITH SASKIA LENAERTS, ALECSANDER ROTHSCHILD, ALEX WOLFE & DINGYUN ZHANG

## WHAT DO YOU WANT A MEANING FOR? LIFE IS A DESIRE, NOT MEANING.

## INFORMATION IS NOT KNOWLEDGE

## COMMON SENSE IS THE COLLECTION OF PREJUDICES ACQUIRED BY AGE 18

## LOUIS VUITTON FASHION EYE KYOTO by MAYUMI HOSOKAWA

## FENDI REFLECTIONS

## FENDI PEEKABOO ISEEU

## FENDI TIMEPIECES SELLERIA MAN

---

cover photo_Yume Ippei fashion_Nao Koyabu grooming_Shuco @3RD
models_Cookie! from YASEIBAKUDAN @YOSHIMOTO & Katsumi Kawahara from TENJIKUNEZUMI + Kenji Yamauchi from KAMAITACHI @YOSHIMOTO

FROM RIGHT: jacket & pants **LITTLEBIG** shirt **DRESSEDUNDRESSED** hat **KENZO** boots **AMI ALEXANDRE MATTIUSSI**
skirt **FACETASM** bra top worn as necklace **DRESSEDUNDRESSED** shoes **PILLINGS**
shirt & socks **STYLIST'S OWN** nasubi & sunglasses **MODEL'S OWN**
jumpsuit & crochet worn as headpiece **PILLINGS**

Addict
CONVERSE

CONVERSE INFORMATION CENTER 0120-819-217

# REMEMBER, YOU CAN ALWAYS STOOP AND PICK UP NOTHING.

all items by **PHENOMENON**

commons&sense man ISSUE09
**photo_Shin Okishima**

1990年代に芽生えたストリートスタイル。2010年代半ばを境に、その機能性とリアリティを体現した普遍的なデザインの中に新たなクリエーションの源を見出したハイブランドも堂々とそのエッセンスを取り入れるようになった。彼らがインスピレーションに求めたのは、紛れもなく東京のストリート。その東京で、いち早くストリートをモードに昇華した立役者こそ、PHENOMENONであった。創業者のオオスミタケシは、ヒップホップユニットのシャカゾンビとして音楽活動を行なう傍ら、1999年にユニットメンバーの井口秀浩とともにSWAGGERを立ち上げてファッションシーンに参入。2004年、オオスミ自身のソロプロジェクトとしてローンチしたのがPHENOMENONだ。2010年3月の東京コレクションでランウェイデビューを果たしたが、リアリティの強いストリートの要素をクリエイティブに洗練させた元祖"ストリートモード"のコレクションは、東京を超え、世界のファッションシーンを震撼させた。その才能に魅せられたcommons&sense manでは、ISSUE09（2010-2011 AW）にてインタビューを敢行している。"ストリートモード"という切り口とともに時代の寵児となり、周囲からもてはやされていたにもかかわらず、本人は飄々として「（ストリート、モードという認識は）まったくありません」と言い切った。それは謙遜か、それとも虚勢の現れか。いずれにしても、世の中を俯瞰する冷静な視線が彼の美意識の根源にあったことは間違いない。2021年2月、突然オオスミの訃報が届いた。享年47歳。まさにこれからという、油の乗り切った年齢であった。今でも、巨大な体躯に似合わぬ愛くるしい微笑みが脳裏に焼き付いている。東京のモードシーンを牽引し、世界のファッションにも影響を与えたオオスミタケシの

**Street style fashion that emerged in the 1990s developed into a well-balanced mix of elements of functionality and realism, essences that would be incorporated by countless luxury brands from the mid-2010s on. Their inspiration undoubtedly came from the streets of Tokyo. PHENOMENON was one of the first brands to sublimate Tokyo street fashion into luxury. Takeshi Osumi, the founder of the PHENOMENON, entered the world of fashion in 1999 when he established the brand SWAGGER with Hidehiro Iguchi, who also performed along with in their hip-hop unit SHAKKAZOMBIE. In 2004, Osumi established PHENOMENON as a solo project. He made his runway debut at Tokyo Fashion Week in March 2010, and his unique way of merging streetwear and luxury creatively refined everyday streetwear elements shocked not only the Tokyo fashion scene but also the at an international level. Fascinated by his talent, our team at commons& sense man conducted an interview with Osumi in our ISSUE09 (Autumn Winter 2010-2011 issue). From our perspective, he was looked up to by many as being ahead of his time in the way he challenged the use of streetwear and luxury, although he nonchalantly stated that he was not consciously mixing streetwear or luxury at the time. He could have just been modest or even pretentious, but nonetheless, there is no doubt that his calm, and big-picture perspective of the world was at the core of his style. News of Osumi's passing came suddenly in February 2021. He was only 47, at the prime of his career. His lovable smile juxtaposing his large physique continues to burn in the back of our mind. We'd like to look back on the trajectory of PHENOMENON in honour of Osumi, who led the Tokyo fashion scene and**

commons&sense man ISSUE07

photo_Akihito Igarashi fashion_Shinichi Mikawa grooming_Go Utsugi model_Kenny N

commons&sense man ISSUE09

photo_Takuya Uchiyama fashion_Shinichi Mikawa hair_Go Utsugi make up_Hisano Komine

models_Hasen, Mao, Nozomu, Robert, Atsuro Mukai, Nemo, Toru Umeda & Yutaka Idota

photos_Hyunseong Kim fashion_Shinichi Mikawa hair_Seungwon Kim make up_Hyeryung Park models_Myungil Lee & Gunwoong Kim coordination_Myonghee Kim

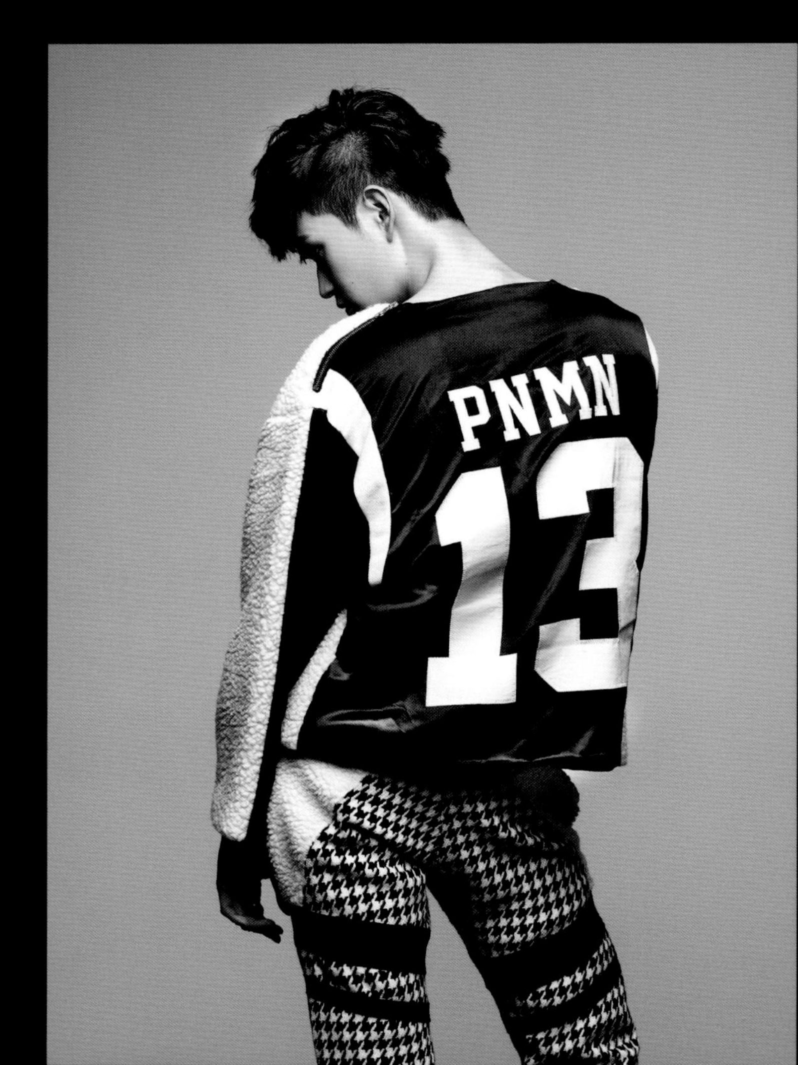

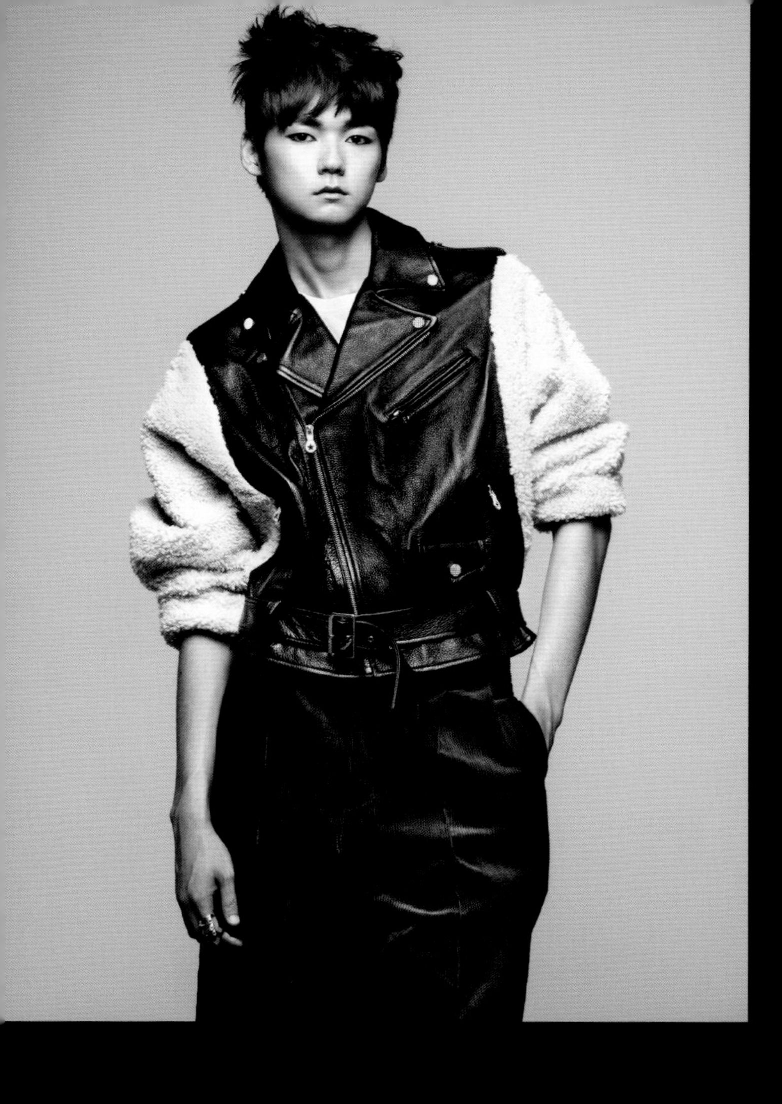

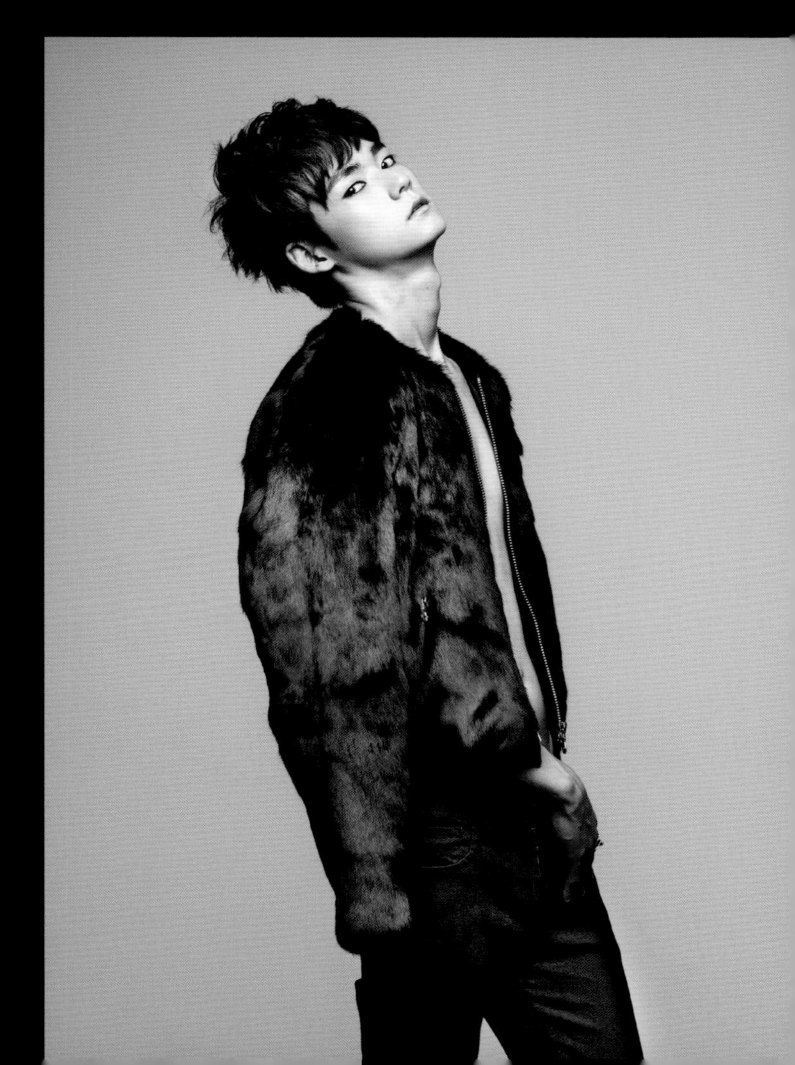

何と生き辛い世の中になってしまったのだろう。声高に叫ばれるコンプライアンス。少しの不祥事も許さない不寛容なオーディエンス。追い討ちをかけるように、新型コロナウイルスが蔓延し、外出さえままならなくなった。巷には“自粛警察”が蔓延り、他人の外出にうるさく口を挟む。日本人特有の同調圧力、匿名の無責任な発言や非難めいた視線につねに怯えて暮らす日々のストレスたるや、尋常ではない。そんなときこそ人びとを楽しませるべきエンターテインメント、特にテレビの世界にも、この悪しき気配は蔓延しているようだ。視聴者にこびるコンテンツ作り。予定調和の結末。実に嘆かわしい。そしてとうとう、芸能最後の砦であった笑いの世界にも、“それ”は侵食しはじめた。本来笑いとは、反骨精神や風刺の姿勢で社会に風穴を開けるべき存在なのに。アレを言ったらダメ、コレをやったらダメ。それでは厳しい社会を吹き飛ばす、爆発力のある笑いなんて生まれはしない。結果、毒にも薬にもならない、ヌルい笑いの番組が主流になりつつある。腹の底から笑うことさえできない世の中に、一体どんな楽しみがあるというのか…… と、

What a difficult world we live in now. All the talk about compliance, unforgiving audiences that won't let any misconduct slide, and to top it all off, the COVID-19 pandemic making it impossible for us to even leave the house. The "quarantine patrol" civilians have made it hard to even go outside. The daily stress of living in constant fear of strangers' snarky comments, accusatory glances, and the kind of peer pressure peculiar to Japanese culture is unbelievable. At times like this, the entertainment industry, especially television, which exists for the sole purpose of entertainment, is unfortunately hit by these bad vibes. It's a shame to have to see television content creators tip-toeing around the aggressively opinionated viewers. And now, the mood has started to eat up comedy, too, our only remaining hope of the entertainment industry. In a normal world, comedy is something that is rebellious in spirit with a satirical attitude, but if you take that away by saying you can't say this, you can't do that, what is left of comedy to make us laugh away all of the society's worries and problems? As a result, what's become of mainstream comedy is lukewarm laughter that is harmless but pointless. What is there to life if comedy can't make

# A DAY WITHOUT LAUG

愚痴を並べ立てたところでフと気付く。本当に、今の笑いはイケてないのか？ いや、つい先日、電車の中で悶絶させられた記憶が。周囲に悟られまいと堪えるのに必死だったじゃないか。昨日の夜だって、腹を抱えて笑わせられた。それはいずれも、もはや行き場を見失ったかのようにも見える今の日本の笑いに、見たこともない奇天烈な武器を掲げて果敢に切り込む猛者たちの仕業であった。世間の声などどこ吹く風。たとえ客にドン引きされようとも、たった1人、いや相方さえ、いやいや自分さえ“オモロい”と思えればいいのだとばかりに。諸刃の剣のような尖り切った笑いを繰り出すさまは神々しく、もはや創造の境地である。そこで今号のcommons&sense manでは、己の信じる“笑い”を追求する3人の芸人をピックアップ。それぞれ毛色は異なるが、いずれもテレビという枠組みに固執せず、劇場はもちろん、YouTubeでもネタを披露し、さらにはアーティストや作家など幅広い分野で活動する真の芸人である。筆頭は、野性爆弾のくっきー！ だ。その思考回路は複雑怪

you laugh your face off? ... and as we vent out here, we also wonder. Is the current comedy scene really that bad? No, just the other day, we were in agony on the train, struggling to hold in laughter. Even last night, comedy made us cry with laughter. The particular skits that made us burst out laughing were by fierce comedians who boldly used never-seen-before comedic weapons in the current Japanese world of comedy that seemingly lost its way. They aren't controlled by what the audience thinks. Even if the crowd is disgusted, as long as one person, or just his partner or even only himself found it funny, that was all that mattered. The way they unleash their sharp jokes like a double-edged sword are divine works of art. For this issue of commons&sense man, we spoke to three comedians who pursue their unique way of comedy. They each have a different style, yet are true comedians who do not stick to only conventional TV shows, but also perform skits in theatres and on YouTube, all while branching out to different fields of work such as art and writing. The first interview is with Cookie! from YASEI BAKUDAN. His thought processes are complex and

奇。ときに視聴者が理解に苦しむような独特の世界観を"押し付け"てくるさまも、1周回って面白い。くっきー! ワールド炸裂のアート作品は社会現象を巻き起こし、展覧会を開けば満員御礼。白塗り顔マネはあまりの人気に書籍化された。その他にも数々の著書を発行し、サブカル界隈をも騒がせている。インタビューでは歯に衣着せぬもの言いや掛け合いの中に地頭の良さを滲ませ、破天荒なイメージの下にとんでもない知性が隠されているのではと錯覚、いや実感させられたほどだ。続くは天竺鼠の川原克己。究極の"媚びない笑い"はしばしば相方の瀬下 豊をも困惑させるが、それが一層の笑いを誘う。言葉を必要としない、まるで絵のようなコントは、もはや芸術のレベル。海外でも通用するに違いない。趣味の絵本集めが高じて、ついには絵本作家としてデビューしたが、そこにも型破りのセンスが充満している。それはインタビューでの発言にも波及しているが、彼の言葉を額面通りに受け取ってはいけない。すべては異次元に生きる彼の、奇想天外な発想から生まれたものなのだから。なんて勝手に解釈してみたけれど、間違ってたらごめんなさい。そして最後を飾るのは、かまいたちの山内健司。飛ぶ鳥を落とす勢いの活躍ぶりを見せているが、本人は昔から何ひとつ変わっていない。時折見せる狂気の目つき、サイコパスな思考から生まれる常軌を逸した行動や発言。客席から悲鳴があがったかと思いきや、次の瞬間にそれは爆笑へと転じてしまう。"ゾーン"に入った彼の無双ぶりは、誰もが認めるところである。最近ではコンビのことやプライベートを綴った軽妙洒脱なエッセイ集『寝苦しい夜の猫』を上梓し、作家デビューを果たしている。この3人、みな40代で油の乗り切った芸人なのに、いまだ若手のようなフレッシュ感が漂うのはなぜだろう。そうか。みんな少年の心のまま、肉体だけが年を重ねているのか。幼い頃に膨らませた自由奔放な妄想は、今も彼らの頭の中でますます拡大し続けているのだろう。3本のインタビューを通して、彼らの魅力に少しでも触れていただければ、これ幸い。そうそう、3人とも無類の猫好きだってこと、知ってま

bizarre, and the way he forces his unique perspective on his audience carries its own sense of humour. His super unique artworks have become a social phenomenon, with all of his exhibitions jam-packed with fans. His white-painted face mimicry skit was so popular that it was published as a book. He has also published a number of books that have caused a stir in the subculture world. During our interview, he showed his intelligence with casual conversation and dialogue, which made us realise his level of cleverness hidden under the lunatic mask he wears. The second interview is with Katsumi Kawahara of TENJIKUNEZUMI. His comedic ways that refuse to butter up the audience often baffles his partner Yutaka Seshita, which makes him that much funnier. His comedy is like a work of art created with words, and there's no doubt he will find international success, as well. His hobby of collecting children's books led him to launch his own children's book filled with his unconventional sense of humour. His personality is reflected in his interviews, but you can't take his words for what they are. We've taken the liberty of assuming that everything he says comes from his outlandish ideas because he lives in another dimension, but we could be wrong. In that case, my bad. The last interview was with Kenji Yamauchi of KAMAITACHI. He's been very successful, but that doesn't seem to have changed him a bit. His occasional crazy eyes and outrageous remarks that come from his psychopathic tendencies make some of the audience scream, but in the next moment, he turns it into laughter. Everyone recognises his unbelievable performance level when he is in his zone. He recently made his debut as a writer when he published his light-hearted and stylish essay book "Negurushii Yoru no Neko," about his comedic duo and personal life. Why is it that these three comedians, all in their forties and widely successful, still seem as fresh as young comedians starting out? Now, we realise that they are all still little kids in their minds with adult bodies. The free-spirited fantasies they developed in their childhood have just continued on to blossom. We hope you can experience a glimpse of their charm through these three interviews. By the way, did you know that all three of them are unparalleled cat lovers?

HTER IS A DAY WASTE

# A LUMP OF MEAT. I GUESS.

## interview with COOKIE!

photos_Yume Ippei fashion_Nao Koyabu grooming_Shuco @3RD
model_Cookie! from YASEIBAKUDAN @YOSHIMOTO
fashion assistants_Mayuko Saito & Yuri Nosho
grooming assistant_Takako Koizumi

DO YOU WANT TO SEE MORE ?

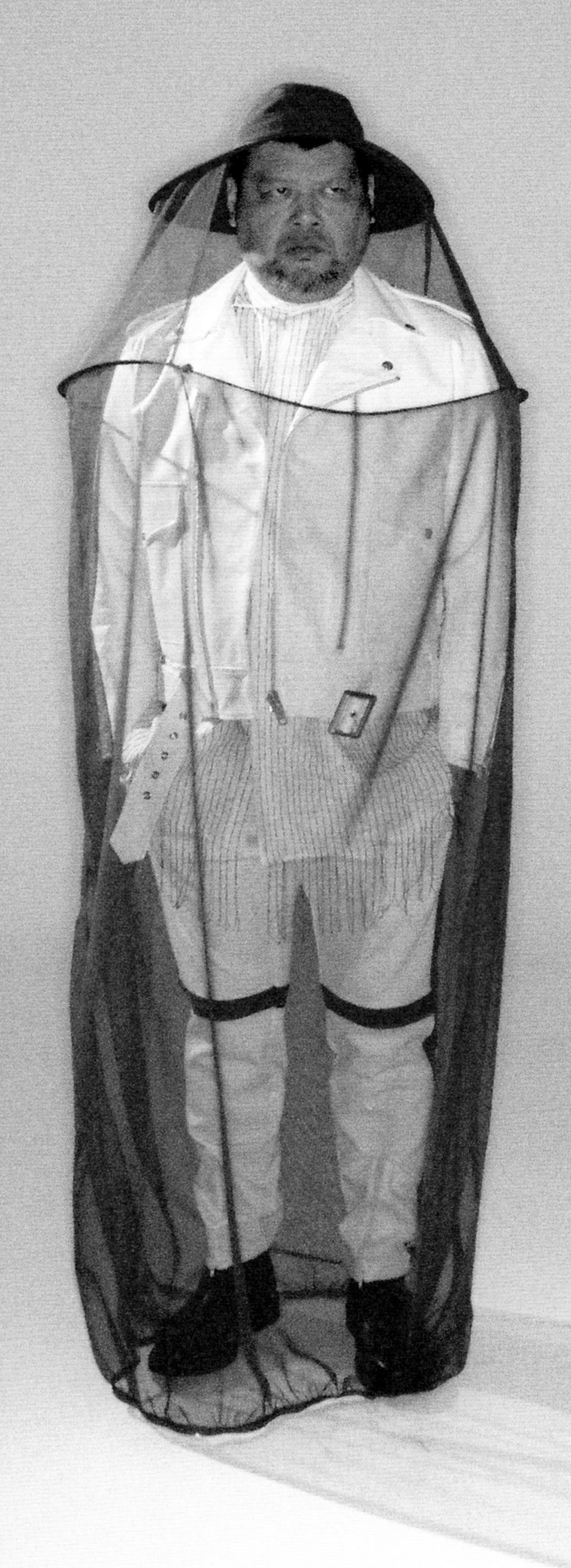

jacket & pants LITTLEBIG  shirt DRESSEDUNDRESSED  hat KENZO  boots AMI ALEXANDRE MATTIUSSI

jacket **LITTLEBIG**

shirt **DRESSEDUNDRESSED**

hat **KENZO**

**commons&sense man (以下CM): お笑いの世界に入ったきっかけは?**
Cookie! (以下C): ツレに誘われた、っていうだけの話なんです。別にやることもないし、(吉本の)養成所行こうって誘われて、全然ええよって感じで。

**CM: 養成所に入る前から、周囲では面白いと言われていた?**
C: それは、まぁ……。1番面白かったんとちゃいますか?(笑)

**CM: どういう"お笑い"を目指している?**
C: 今やってて、なんとなく思ってんのは、"気持ち悪い"と"おもろい"のちょうどええ……間の子っちゅうんか、合体がいいなと思ってて。うわ〜!って目を背けたくなるけどおもろい、っていうのを目指してるとこですかね。お笑いって、やっぱり誰もやってないところをやるっていうのが基本なんで。最初っからこの感じでやってるから、もう麻痺してるというか。痺れ切れた状態でずっと動いてるみたいな、よう分からんまんま。

**CM: 観客より、芸人仲間からの方が評価が高い?**
C: そうですね。っていうか、ほぼそれですね。だから全然怖くもなくて、自分がおもろいと思ってる先輩らが笑(わろ)てくれてるんで。客のこと、おもろいと思ったこと1回もないですし(笑)。

**CM: 売れたいという気持ちは?**
C: ないです。まぁ、売れた方がええ、っていう感じじゃないですかね。やりたいことをやって売れたら最高やな、っていうことの連続です。「売れたい!」やと、売れるために自我を捨ててやらなアカンと思うんですけど、それはやってないんで。売れたらええな、くらい。今んとこ、ラッキーですね。売れてない苦労時代もえらい長くて、芸歴26年なんですけど、多分20年以上はかかってますよ、ちゃんと食えるようになるまで。

**CM: 大阪では観客にもウケていた?**
C: 身内だけね。一般客からしたら、ちょっとクレイジーなのがおる、みたいな感じやったと思います。デビューして、単独イベント、自分らのネタイベントとかやるんですけど、男子率が異様に高い。黄色い声援が基本的にあんまりないというか……(笑)。

**CM: 黄色い声援は欲しい?**
C: 欲しいですよ! そりゃ、ワーキャー言われてぇっすよ(笑)。もう諦めてますけど。

**CM: そのために芸風を変えることはない?**
C: やっぱり、ついて来てくれはったお客さんを裏切るわけにもいかないんで。一部、ササってる女性のお客さんもいますよ、変な人多いですけど(笑)。そんなとこすんねや! みたいなとこにピアス入れてる人もいてました。

**CM: お笑いにおいて必要不可欠なことは?**
C:「折れず、曲がらず」ですかね、基本的には。オンエア見たら、勝手に折れた状態にされてることはありましたけど(笑)。編集されたり、カットされたりで。ただ、今んとこは貫いてやれてるんで、まぁやりたいことをやれてる方じゃないですかね。でも、基本的にオンエアはあんまり見ないです、イヤな気持ちになるだけなんで。たまに自分がメインのゲストで出させてもろてるのに、全発言カット、みたいな爆裂なときがあるんですよ。絶好調にええときのオンエアは見ますけど(笑)。あ、ちゃんと使(つこ)てくれてんねや、と。爆跳ねしてるときとかは、やっぱり使われてるかなぁ。まぁ発言のレベルにもよりますけど。

**CM: "お笑いの神"が降りてきたと思った瞬間は?**
C: あります、あります。僕の持論で、後輩にもよう言うてんのは、でっけぇ番組、メジャーな番組で、3連チャンで跳ねたら売れる、っていう。なんかね、あるんですよ、モード的にノリノリな期間っていうのが。脳みそフル回転で、ブドウ糖満タン、みたいな。そのときにデカい番組に当たったら……。それが運なんかもしれないですけど。僕の場合は、それが2回ありました。大阪時代は、まず『オールザッツ漫才』っていう番組と、『ガキ使(ダウンタウンのガキの使いやあらへんで!)』の"七変化"っていうコーナー。そのあと『アメトーーク!』の"東京にハマってない芸人"の3タテで跳ねて。それで一気に仕事増えましたね。ゴールデンタイムに慣れてないんで、歪な発言ばっかりして、みるみる内に仕事なくなりましたけど(笑)。

**commons&sense man (hereafter CM): What got you into comedy?**
Cookie! (hereafter C): A friend invited me to try out for YOSHIMOTO's NSC (New Star Creation) training school, and since I didn't have anything better to do, I was like, "Sure, why not?"

**CM: Were you the class clown type before getting into the training school?**
C: Well, I was probably the funniest one around (laughs).

**CM: What kind of comedy do you strive for?**
C: What I'd like to achieve is the middle ground of gross out and funny humour. Stuff that makes you want to cover your ears, but you can't because it's hilarious. Comedy is about doing something no one else has ever done. I've been doing it this way from the beginning, so I guess I've become numb to the idea. It's like I've been subconsciously doing it for years without knowing.

**CM: Do you think your fellow comedians appreciate your humour more than the audience?**
C: Definitely. Senior comedians that I respect and find funny laugh at my jokes, so that's been a confidence booster. I've never found the audience to be funny (laughs).

**CM: Do you feel the need to be successful?**
C: Not really. It's better to be successful than not, but for me, it's best if I can be successful doing what I love. If the desire to be successful comes first, you can rarely do something you're proud of, so for me, I just hope that success comes along if I keep doing my thing. I've been lucky so far but the time I've spent under the radar was so long — it took me more than 20 years out of my 26 year career as a comedian for me to even make a living.

**CM: How did the Osaka audiences perceive your comedy?**
C: People that I know loved it. As for the general crowd, they probably just thought I was crazy. Ever since my debut, I've done solo events and other shows, and I've noticed the percentage of men in the audience is much higher than women. There aren't very many girls cheering me on in the crowd (laughs).

**CM: Do you want more fans that are women?**
C: Of course I do! I'd love to be popular among women (laughs), but I've practically given up on that.

**CM: Would you ever change your style to cater to women?**
C: I can't betray my loyal fans. I do have a number of fans who are women although they're mostly out-there in terms of style (laughs). Like those with piercings in places I didn't even know you could pierce.

**CM: What is essential in comedy?**
C: To not break or bend yourself. There were times when I would see my performances on TV edited or cut in a way that makes me look broken (laughs). But so far I've been able to stick to my guns and do things the way I want to. I try not to watch myself on TV because the editing makes me feel bad about myself. There've been times where I was the main guest of a show but all of my comments got edited out. I do watch some shows where I did great and they used my cuts, though (laughs). They do usually keep the scenes where I get a great laugh from the audience, but I guess it depends on how funny it was.

**CM: Was there ever a moment when you felt the power of a comedy God?**
C: Yes, definitely. I have a theory, and I tell this to younger comedians too, that if you are on a big major show and are able to make the crowd burst out into laughter three times in a row, you will be successful. We all have certain periods of times when you're in the groove and your brain is in full swing and your body is full of glucose. If you are able to be in a major TV show during that brief moment, it may be luck, but for me, it's happened twice. The first was when I was in Osaka in a show called "All That's Manzai" and the second was on the show "Gaki-Tsuka (Downtown's Gaki no Tsukai ya Arahende!)" in the Seven Changes scene. After that, I was on "Ame-talk!" in a program where they featured comedians who don't fit into Tokyo. After that, I suddenly started to get more job offers. I wasn't used to being on primetime TV, so I made a lot of inappropriate comments which resulted in a decline in the offers very quickly (laughs).

CM: インスピレーションの源は?
C: どうなんですかね……。お笑いも、それ以外の活動も、発想は一緒やと思います。もともとお笑いをそんなに熱心に見てた方でもなくて、誘われてこの世界に入って来たくらいなんで。ガキの頃に見てたのは、バンドとか映画とかじゃないですかね。映画も昔のやつばっかりで、伊丹十三さんの作品が好きでした。あとは『ビー バップ ハイスクール』、『バック トゥ ザ フューチャー』とかジャッキー チェン系とか。その辺ちゃいますかね。でも今、映画はあんまり見ないんですよ。だから更新されてなくて、映画の脳みそが。バンドは、バンドブームの後半くらいからなんで、ジュンスカ(JUN SKY WALKER(S))とかCOBRAとか。中学時代、暴走族の先輩と1回喧嘩になっちゃって。ボッコボコにされた後に「お前、音楽何聞いてんねん」って訊かれて「ジュンスカです」って答えたら、「明日俺んちのポストにジュンスカのCD全部入れとけ」って言われて。目ぇパンパンに腫らしながら、CD持って行ったん覚えてますもん、中2のとき。あれは地獄でした、ホンマ(笑)。そのあと仲良くなったとかもないです、ただボッコボコにされただけ(笑)。

CM: 朝起きて1番にすることは?
C: 朝イチですか?タバコじゃないですかね。若い頃、1回タバコの銘柄を変えたいなと思ったことがあって。タバコ屋のオバハンが言うには、黄色の"ロングピース"は殺人鬼が吸うタバコやと。匂いがきつい分、血の匂いも消してくれるから、殺人鬼はみんなこれ吸うんよと言われて「カッコええな、おばちゃん、ほなピース頂戴!」って。で、ずっとピースの時期ありましたね(笑)。めちゃめちゃニコチンもタールもキツいんすよ。で、朝イチに寝そべって吸ってたときに、キツすぎてフラフラんなって起き上がられへんようになって、やめましたけど。

CM: 今1番幸せを感じるときは?
C: ガキの頃欲しかったもんって、今でもやっぱり欲しいじゃないですか。それを惜しみなく買えた瞬間。金払ってる瞬間、やっぱり気持ちいいっすね。1番最近やと、まだ届いてないですけど、それで単車買いました。トライアンフ。あとはギターですかね。ギターはわんさか買ってます、20本くらい。自分でも弾きますけど、男ってコンプリート癖、あるじゃないですか。やっぱりキン消し(キン肉マン消しゴム)のせいやと思うんですけど。キン消し全種欲しかったっていうあの欲は、今でも残ってる。だから、何でもつい全種欲しなってしまうんですよねぇ。

CM: 今もっとも興味があることは?
C: 御朱印集めですかね(笑)。3年前くらいにはじめて、御朱印帳もちゃんと買(こ)うて。で、まだ2個。今んとこ2朱印だけです、2神社(笑)。でも、わざわざ持って歩かないじゃないですか。で、金払(はろ)て字書いてもろてる間、待っとかなあかんし、めんどくさい。神社自体も好きなんで、最寄りの氏神さんの神社には、初詣等々、ちゃんと行きますよ。

CM: 意外に信心深い?
C: どうなんでしょうね、でも信心深かったらこんな気持ち悪いネタやらへんと思いますよ(笑)。

CM: 尊敬する人は?
C: いっぱいいますよ。お世話になった人は、やっぱり尊敬の対象ですし。たとえばバッファロー吾郎さん。あと、ハリウッドザコシショウさん、ケンコバさんとか。僕がデビューしたてのときから、なんなとお世話になってるんで。バッファローさんは、コンビなんですけど、今は各々で出てはります。今でもやっぱり面白いですね。雑誌とかで、バッファロー吾郎さん尊敬してる言うたら、FUJIWARAのフジモンさんに、バッファローさんと同期やからか、ちょいちょいキレられんすよ。バッファローばっかり名前出すな、言うて。でも、フジモンさんを尊敬することはできないっすね(笑)。めちゃめちゃ仲良くてええ人なんですけど、尊敬はゼロです、無です。

CM: この世で1番おもしろいと思う人は?
C: (ハリウッド)ザコシショウですね。桁違いというか、規格外なんで。目鼻立ち、めちゃめちゃキレイで男前なんですよ。なのに、あの体型。神さまの悪戯でしかないと言うか。豚ホルモンみたいな体してるじゃないですか。もうその時点で面白くて。ある意味、吉本離れて良かったかもしれないですね。吉本に居続けてたら、下手したら、地下ではオモロい人、っていうだけで終わってたかもしれないですし。

**CM: Where do you get your inspiration?**
C: That's a hard question... I think comedy and other activities come from the same idea. I was never an avid comedy fan growing up, and I only got into comedy because I was suggested to by a friend. I grew up watching band performances and films. I liked old movies directed by Juzo Itami, "Be-Bop High School", "Back to the Future", and Jackie Chan movies. I don't watch that many movies, even, so my knowledge of movies stops there. I've been into bands since the latter half of the band boom, like Jun Sky Walkers and Cobra. In junior high, I got in fight with an older biker gang kid. He beat me up and asked me what kind of music I listened to, and when I answered JUN SKY WALKER(S), he told me to put all my JUN SKY WALKER(S) CDs in his mailbox the next day. It was hell, eighth grade me walking up to his house with swelled up eyes with my treasured CDs (laughs). It wasn't even like we became good friends after, it's just a tragic story. (laughs)

**CM: What's the first thing you do in the morning?**
C: The first thing I do is smoke, I think. Back in the day, I changed up my brand of cigarettes after the tobacco shop old lady told me that the yellow "Long Peace" cigarettes were for serial killers because the strong smell removes the smell of blood. (laughs). It had strong nicotine and tar levels, and one time I was smoking it lying down first thing in the morning, and it hit so hard that I got dizzy and couldn't get up. That's when I quit smoking the yellow "Long Peace".

**CM: When do you feel the happiest?**
C: When I can afford buying things that I wanted as a kid, and still want now, you know? The moment when I pay without hesitation. The most recent purchase, although I haven't received it yet, is a Triumph motorcycle. Another thing I bought was a guitar. I've bought a lot of guitars, about 20 of them. I play them myself, but I think men have a tendency to collect things. I think it comes from wanting to collect all of the "Kinkeshi (Kinnikuman erasers) " as a kid. I still have that desire to collect things in sets.

**CM: What are you most interested in now?**
C: I started collecting shuin (a seal stamp given to visitors to shrines and temples) about three years ago, and I bought shuin books. So far, I've only collected from two shrines (laughs). The thing is, it's a hassle carrying the book around, paying to get the stamp, and waiting for them to write calligraphy in the book. I do like Shinto shrines, so I go to my nearest shrine to pay my respects for certain occasions like New Years.

**CM: Are you religious?**
C: I don't know, but if I were religious, I don't think I would be performing my repulsive jokes.

**CM: Who do you look up to?**
C: There are so many people I look up to. I respect people who have helped me along the way. Buffalo Goro, Hollywood-Zakoshisyoh and Kenkoba, to name a few, have been kind to me since I was just starting out. Buffalo used to be a comedy duo, but now they work individually. They're still funny even now. When I tell people in magazines that I respect Buffalo Goro, Fujimon from FUJIWARA gets mad at me because he's the same age as he is. He tells me to stop mentioning Buffalo (laughs), but I can't respect Fujimon the way I respect Buffalo. We get along really well and he's a great guy, but I have zero respect for him.

**CM: Who do you think is the most funny person in the world?**
C: Definitely Hollywood-Zakoshisyoh. I mean, he's out of this world. His facial features are beautiful and he's so handsome, and yet, he has that pig innard-like body for God only knows what reason. His mere presence is hilarious. In a way, I'm glad he left YOSHIMOTO. If he had stayed he might have ended up being just an underground funny guy.

**CM: 1番おもしろくない！と思う人は？**
C: 大山英雄ちゃいます？いや、ようそんな質問しますね、怖い質問！

**CM: 1番"賢いな！"と思う人は？**
C: 勉強で言うたら、単純にロザンの宇治原は賢いですけど。（南海キャンディーズの）山里ちゃいますかね？南海がバっと来たとき、しずちゃんばっかり出てたんですよ。それでも腐らず、「しずちゃんしか仕事ないんですよ〜」って、マイナスを生かしてプラスに変えてた。その瞬間じゃなくて、さらに先を見てると言うか。結局そのあとMCやったり、ド真ん中行ってるんで、やっぱりすごいなぁと思いますね。

**CM: 1番"アホやな〜"と思う人は？**
C: 天竺（鼠）の川原かなぁ。アイツは爆発的にオモロいんすよ。みんな天才って言うんですけど、その天才っぷりの生かし方をよくわかってないというか。その場所にあったやり方ってあるじゃないですか。それこそ、イカ釣り漁船乗ってんのに、ヤリでイカを取ろうとしてるヤツ、みたいな。いや、それはやり方ちゃうで、っていう（笑）。すごい天才なんですけど、生き方はアホやなって思います。

**CM: 2人は似ていると言う人もいるが？**
C: いや、嬉しいっすね。仕事一緒にしたい芸人なんで。意外とうまく回るんですよ。よう飯行ったりもしますし。なんとなく、川原の使い方も熟知してるつもりなんで。川原をテレビに呼びたい、でもちょっと怖いなっていう人は、僕一緒に出してもろたら、全然上手に扱いますよ（笑）。川原とザコシショウは、僕、扱い方上手やと思います。

**CM: 今1番恐れているモノは？**
C: 東京って、逆に緑が多いから、案外スズメバチがおるらしくて。スズメバチが1番怖いです……。

**CM: 東京に拠点を移した理由は？**
C: 大阪でメインで活動してた劇場（うめだ花月）が建物取り壊しでなくなって、代わりに京橋に新しい劇場（京橋花月）を作ることになって。うめだ花月では、中堅も含めて若手だけでやってたんですけど、大きい劇場には師匠も出ることになって。そしたら師匠にわざわざ挨拶行かないとダメじゃないですか。それ、ちょっと鬱陶しいなぁって。世話にもなってない知らん師匠のとこに、なんで挨拶行かなあかんねん、言うて（笑）。ほんで、メインの劇場が潰れるんやったら東京行こかと。ロビーでピリっとするんも、イヤじゃないですか。ウケもせん、知らんジジイと（笑）。あ、ちょっと見たことあるくらいのジジイか（笑）。

**CM: お笑い以外の活動を通して表現したいことは？**
C: そんな深く考えてるわけじゃないんですけど、軸はお笑いっていうのはあります。やってくれって言われてやるとか、あとは何となく「これやったらオモロいやろうな」って思ってやってるだけで、「この表現が……」みたいに考えたことはあんまりないですね。

**CM: 自分を一言で表すとしたら？**
C: 肉の塊、ですかね〜。

**CM: 10年後、ご自身はどうなっていたい？**
C: 出た、その質問！これ、僕の持論があって。『シザーハンズ』って言う映画は、バケモンが人のいっぱいおる街中に下りて来て、最初は気持ち悪がられてるけど、だんだん「こいつ可愛いな」ってなってきて、人気出て、でも最後はまた嫌われて山に帰らされるって言う話で。僕の人生もそれやと思ってて。今たぶん、街中で好かれてる状態なんですけど、「大阪からバケモン来た！うわ、気持ちわる……。でも、なんか可愛いで？」言うて。で、そのうち追い返されると思うんで……（笑）。10年後は、山奥の掘立小屋で、1人で正座してるんちゃいますかね。

**CM: Who do you think is the most unfunny?**
C: Probably Hideo Oyama, right? What kind of scary question is that?

**CM: Who do you think is the most clever?**
C: In terms of being book smart, definitely Ujihara from ROZAN, but I think Yamasato from NANKAI CANDIES is clever. When NANKAI CANDIES started appearing on TV, Shizu-chan was the only one who got attention, but he didn't lose faith and turned something negative into a positive thing by saying, "Shizu-chan is the only popular one." He was looking at the big picture, and in the end, he's become a great MC and went on to doing great things, and I think that's amazing.

**CM: Who do you think is the most stupid?**
C: Kawahara from TENJIKUNEZUMI. He's explosively funny. Everyone says he's a genius, but he doesn't really know how to make the most of his genius-ness. There's a time and place for everything, and he just doesn't get it. He's like a guy on a fishing boat trying to catch squid with a lance, like, you're doing it all wrong (laughs). He's a genius, but his way of living is brilliantly dumb.

**CM: Some people say that you two are similar.**
C: That's a great compliment. He's someone I'd like to work with. We get along surprisingly well. We go out for dinner often, too. I think I know how to operate him pretty well by now. If anyone wants to cast Kawahara on a show but are a little scared, you can hire me as a set and I'll tame him (laughs). I think I can definitely handle Kawahara and Zakoshisyoh very well.

**CM: What are you the most afraid of now?**
C: Tokyo has a lot of greenery, so apparently there are a lot of wasps. I'm most afraid of wasps...

**CM: Why did you move your base to Tokyo?**
C: The theatre where I mainly performed in Osaka (Umeda Kagetsu) was demolished and a new theatre (Kyobashi Kagetsu) was built in Kyobashi. At Umeda Kagetsu, we were performing among young and mid-career comedians, but at the new bigger theatre, older comedians were going to be performing, too, which was a bit off-putting for me, because you'd have to go and pay your respects to them at each show. I thought, why would I have to go and say hi to a older comedian I don't even know? (laughs) I thought, if the main theatre was going to shut down, I should just relocate to Tokyo, so I wouldn't have to tense up in the lobby with some unfunny old comedian guy that I didn't even know (laughs).

**CM: What do you want to express through activities other than comedy?**
C: I don't really think about it that much, but my main focus is comedy. I just do what I'm asked to do or I do what I think would be funny, but I don't really think about the deep meaning and expression and whatnot.

**CM: If you had to describe yourself in one word, what would it be?**
C: A lump of meat, I guess.

**CM: Where do you see yourself in 10 years?**
C: There it is, that question! I have a theory about this. In "Edward Scissorhands", he's first treated as a monster in the city and people were scared of him, but they eventually realise he's cute and becomes popular. But in the end they hate him again and he goes back to the mountains. I think that's what my life is like. I'm currently in the phase where I'm well-liked in the city, and I've had people say to me, "You're the monster from Osaka! Ew! But he's kind of cute, isn't he?" I'm sure one day soon they'll turn their backs on me... (laughs). In 10 years, I'll probably be sitting alone in a shack deep in the mountains.

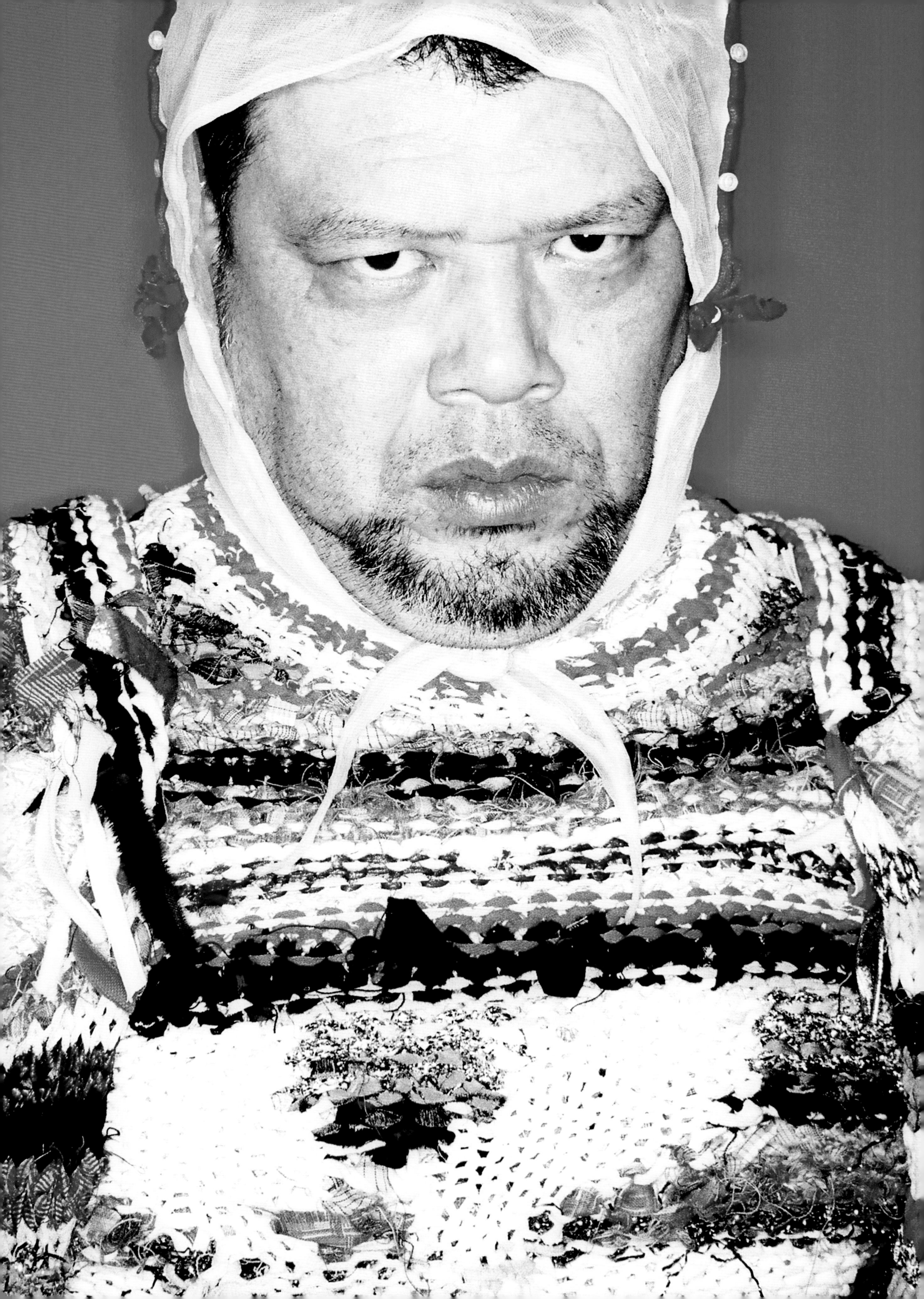

NO, I'M GOING TO SPEND MY WHOLE LIFE CLIMBING THIS DIRTY MOUNTAIN!
interview with KATSUMI KAWAHARA
photos_Yume Ippei fashion_Nao Koyabu grooming_Shuco @3RD
model_Katsumi Kawahara from TENJIKUNEZUMI
fashion assistants_Mayuko Saito & Yuri Nosho
grooming assistant_Takako Koizumi
DO YOU WANT TO SEE MORE ?

skirt **FACETASM**

bra top worn as necklace **DRESSEDUNDRESSED**

shoes **PILLINGS**
shirt & socks **STYLIST'S OWN**

skirt FACETASM

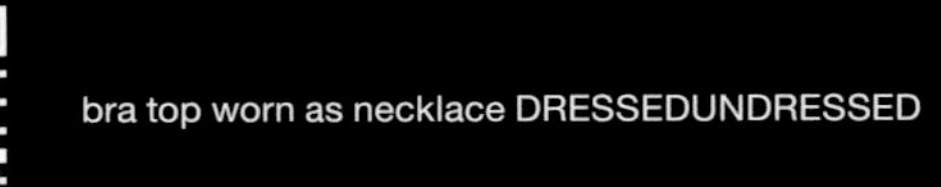

bra top worn as necklace DRESSEDUNDRESSED

shoes PILLINGS
shirt & socks STYLIST'S OWN

commons&sense man（以下CM）：お笑いの世界に入ったきっかけは？
Katsumi Kawahara（以下KK）：それはねぇ…… いつも皆さんにお任せしてるんですよ。

CM: どういう"お笑い"を目指している？
KK: 11歳から88歳までの方々に笑ってもらえるようなお笑いですね。イタリアからベトナムの方まで。

CM: 今の日本の"お笑い"をどう思う？
KK: そんなこと、親にも聞かれたことないですよ!!勘弁してください!!

CM: お笑いにおいて必要不可欠なことは？
KK: …………………… "間（ま）"です。ホンマは今の、半日くらい開けてから"間"って言いたかったのですが今回は、時間がないんで……。僕はテレビの間より舞台の間の方が好きです。ネタを配信したらいいのにってよく言われるんですけど、画面越しに見ている"間"と、実際に現場で見る"間"とは少し違うんです。だから、配信はやりたくないんですよね。で、"無観客無配信"とかやっちゃうんです。本来は1時間くらいで終わる予定が、居心地良すぎて、ホンマに誰も見てないのに（笑）、2時間40分くらいやってました。

CM: "お笑いの神"が降りてきたと思った瞬間は？
KK: 鹿児島でロケしてるときに、霧が出てきて、ひげの長いおじさんみたいなのが挨拶してきたから「誰ですか」って聞いたら「お笑いの神様だ」言うてましたけどね。そのときにしかお会いしたことはないですねぇ。現場でよく言う、奇跡的な笑いをもたらす神さまのことは、ちょっと存じ上げないです。

CM: インスピレーションの源は？
KK: 絵を描くときも、右脳で描くときと左脳で描くときがあって。左脳で描くときは、自分が面白いと思っているものを絵にしようと思って描くんですけど、右脳のときは、何を描こうとか意識せずに、ペンをキャンバスに置いて、動かしていく中で出てくる。ネタも同じで、左脳で考えるときは、この一言を言いたい、このボケを入れたい、それならこんなヤツがやったら面白いんじゃないか、じゃあ学校のコントにしよか、という感じで、内側から作っていく感じ。右脳の場合は、絵と同じで、たとえばまず小道具を作ってみる。お寿司のネタも将棋のネタも、どういうネタにしようとか考えずに、ただ被り物を作ったところから生まれました。まずは被ってみて何か感じて、体が勝手に踊ってたって感じ。右脳と左脳とで、作り方がまったく違いますね。仕上がりも、右脳か左脳かで全然違います。最近は、左脳より右脳の方が多くなってますね、そっちの方が好きなんで。

CM: どこに向けてネタを作っている？
KK: 自分がお客さん側にいたら羨ましいやろなというネタを作ってます。自分で嫉妬しちゃうくらい。たくさんの人に笑ってもらいたいというよりは好きなことをやるのがもうゴールでプラスで誰かが笑ってくれたらラッキー、売れたらラッキー、テレビに出れたらラッキーって感じかな。どうやったら売れるか、テレビに出られるかっていう発想がそもそもないですね。ネタにしても、このままで使ってもらえるなら出ますけど、テレビ向きにちょっと変えて欲しいと言われたら、あ、じゃぁ結構です、と。生意気に、もう1年目からそういうこと言ってました。売れてからでも好きなことはできるって言う人もいますけど、そんなことはない。散々好きでもないことをやったあと、じゃあ好きなことをしてもいいよとなったときには、もう脳みそがついていかなくて、自分が何を面白いと思ってるのかも分からなくなってると思うんです。ただ、僕はホンマに面白いことって何なんやろう、って突き詰めて考えるタイプですけど、お笑いを手段として、テレビで活躍したいっていう人もいる。そこは、同じプロでもまったく視点が違うと思います。なんちゃってね。

CM: 売れている同期をどう見ている？
KK: 素直に嬉しいですね。たとえばかまいたちとかは誇らしくもあります。悔しくないのかとよく聞かれますが登ってる山が違うからね。みんながすごい立派な山に登ってるときに、僕は誰も登ってない、きったない山を1人目指してワクワクしてるんで。みんなにこっちの山に来いって言われても「いや、この汚い山ひとつ登るのに、僕は一生を賭けます!」みたいな。好きな山に登りたいと言う気持ちは、誰もが昔は持ってたと思うんです。僕はそれを今でも持ち続けてるだけ。いまだにそんなこと言ってるのはダサいしカッコつけてるだけって言う人もいますけど、だとしてもそれの何が悪いのかわかりません。ダサいとかカッコいいは人が決めるんじゃなく自分が決めるものですから。人の意見はどうでもいいです。…今のかっこよかったですか？

commons&sense man (hereafter CM) : What got you into comedy?
Katsumi Kawahara (hereafter KK) : I always leave that up to your imagination.

CM: What kind of comedy do you strive for?
KK: The kind of comedy that can be funny to people from 11 to 88 years old, from Italy to Vietnam.

CM: What do you think of the current comedy scene in Japan?
KK: Oh pleeeease! My parents don't even ask me that!

CM: What is essential in comedy?
KK: ………………… "ma (Japanese concept of space and time)". I want to pause for half a day before answering that question, but we don't have that kind of time. I prefer acting on theatre rather than being on television. I'm often told that I should broadcast my own content, but the "ma" on-screen differs from the "ma" on stage, so I don't want to do that. I end up doing something like "no audience, no broadcast." I was just going to do it for an hour, but it got so comfortable that I did it for like 2 hours and 40 minutes, even though no one was watching (laughs).

CM: Was there ever a moment when you felt the power of a comedy God?
KK: When we were on location in Kagoshima, an old man with a long beard appeared from the fog, and when I asked him who he was, he said, "I am the God of Comedy." If you're talking about the God that everyone talks about on set, no, I don't know anything about the man that miraculously brings laughter.

CM: Where do you get your inspiration?
KK: There are times when I draw with my right brain and times when I draw with my left brain. When I'm drawing with my left brain, I try to draw what I find interesting, but when I'm drawing with my right brain, I don't even think about what I'm going to draw, I just put the pen on the canvas and move it around. It's the same with skits. When I think with my left brain, I think about words to say and when to be funny, and things like "Wouldn't it be funny if this guy did that?" and decide to do a school skit. When I use my right brain, it's the same as when I draw. I start making props first without thinking about what kind of skit it will be. My sushi skit and the shogi skit were both born from first putting on a headpiece without thinking about what kind of story it would be. In that case, I'd try on the props and feel, and then think of the jokes later. My right brain and left brain are completely different in the way they give me ideas, and you can see the difference in the result. Recently, I've been using my right brain more than my left brain, because I like how it thinks it better.

CM: Who do you cater your comedy towards?
KK: I make skits that I think would be envious if I were in the audience. Like, I get so jealous of myself. Rather than trying to please everyone in the room, I'm just doing what I love and that is the goal. Plus, I'd be lucky if people would laugh or I'd be lucky if I sell and I get to be on TV. I never think of what I should do to try and get on TV. If TV production people want to use my skits as is, I'd be happy to perform them, but if they want me to change it to make it more TV-friendly, I'll just say no. I know it's audacious of me, but that's how I've been doing it since day one. Some people say that you can do whatever you want after you become successful, but I don't think that's true. If you do all these things you don't want to do in order to become famous, your brain won't be able to handle something interesting once you become famous. I'm the type of person who goes after what is truly funny, but there are people who become comedians for the sake of becoming famous on TV. Even as the same industry professionals, these are completely different perspectives… just kidding.

CM: How do you view successful comedians that started around the same time as you?
KK: Simply happy. For example, I am so proud of KAMAITACHI. People ofthen ask me if I envy them, but the mountains they climb are different than mine. While everyone is climbing great mountains, I'm excited to be climbing alone on a trivial mountain that no one else has climbed. When people tell me to come over and climb their mountain together, I'm like, "No, I'm going to spend my whole life climbing this dirty mountain!" I think everyone at one point had the desire to climb their favourite mountains, but I'm just still doing that. Some people think it's tacky, but I don't see any problems with that. Whether you're tacky or not, should not be judged by other people, but yourself. I don't care what other people say… Did I sound cool?

**CM: 小さいときはどんな子どもだった?**

KK: 小中学校では自分で好んで孤立してたとこがあって、クラスの中心的なグループがチョケてワーワーやってても、僕は教室の端っこで「ちゃうちゃう、今お前が喋ったらアカン、あ、今はお前が動かんと!」とか、ずっと妄想しながら見てました。だから、自分で何か言ったりやったりして人を笑かしたことはないんです。で、高校のときに、野球部で仲の良かった先輩が「文化祭でお笑いするから台本見てくれ」と。この先輩だけは、僕がボソっという一言とかを聞いて「お前、面白いな」と言ってくれてた人なんでアドバイスを求めてきたんです。それで僕は、その台本みてすぐに「あ、面白くないですね」と(笑)。それから具体的にここがこうで、ってニュアンスを伝えたら、「ちょっとわからんから、お前出てくれ」って言われて、3年生の文化祭の出し物に出て。それで周りから面白いと言われるようになりました。それからというもの、みんなが僕を輪の中心にしようしようとしてきたのですが目立つのは興味なかったので1人で妄想してた方が楽しかったですね。

**CM: 東京に進出した理由は?**

KK: 普通のインタビューでは、大阪よりセンターマイクの数が多いって聞いたから、って答えてますけど(笑)、今日は先ほど、美味しいパンを頂いたのでちゃんと答えましょう。単純に、見てくれる人とか、ネタをやれる場所が広がるっていうのがありましたね。吉本には、賞レースで決勝に行ったり優勝してから東京に出るっていう流れがあるんですけど、それはイヤでした。どうせなら1番変なタイミングで行きたいなと。だから「え、東京来てたん?」っていう反応、多いですよ。

**CM: お笑いの世界で生きていこう、と決めたのはいつ頃?**

KK: 今、たまたま芸人をやってるだけで、ここがゴールなわけではないんです。高校卒業してすぐにお笑いの世界に入らなかったのは、まず就職してみたかったから。お笑いの世界に入ったら普通の仕事の経験できなくなるから普通に働くなら今しかない! と。自分はどこまでできて、どこまで認められるのかを知りたくて、最初は大工をやって、大手の工務店に入ってみたりとか、色々やってみて。うなぎ工場とか市場、結婚式場でも働きました。うなぎ工場と市場では、正社員にならないかとか次の社長にならないかって言われたんですけど、そうやって認められたら「俺、ここまでできたんや!」って満足して、その次の日に辞めたりするんです。芸人も、好きなことがある程度できるようになって満足したときには、やめてもいいかなって。その次に何をやるとかは、別に……。自分を知るためにいろいろやってるだけなんで。ただ、1番好きなのはお笑いなんで、それを軸として人生を考えてはいます。

**CM: この世で1番おもしろいと思う人は?**

KK: もう、人間全員です。たとえば、タワーマンションの1番上の部屋で、横になっている人がいると思うと……(笑)。あんな高いところなのに、いつの間にか寝ちゃってたとか……。飛行機でも、上空を飛んでるところをスケルトンで見たい。あんな高いところで映画見てるとか、絶対面白くないですか? マスクにしても、みんな普通に耳に掛けてますけど、元々耳ってものを掛けるところじゃない。なのに、当たり前のように掛けてるのが、なんか可愛いし、面白いなと思います。だから、全員面白い。

**CM: 1番おもしろくないと思う人は?**

KK: タバコを吸ったらアカンところで吸ってる人ですかね。あと、地球に優しくない人とか、否定を簡単にする人も面白くない。お笑いは単純に面白い、面白くないで評価されますけど、それ自体、美術館の絵を見て、ヘタか上手いかだけで決めるのと同じこと。そういう人は美術館を5分くらいで出ちゃうんだろうなと思ったら、やっぱり面白くないですね。

**CM: 1番"賢いな!"と思う人は?**

KK: エジソンです。

**CM: 1番"アホやな〜"と思う人は?**

KK: やっぱりエジソンですね。自分の工場が燃えたときに、息子を呼んで「滅多に見れないから見とけ」って。すごくないですか? これまでの資料とか全部なくなってるのに「これでまた新しいことが考えられる!」って。賢いのに、ほんまにアホやなって。ただただ、チャクラがダダ開きの人やと思うんです。もう、(頭の中が)子どものまま大きくなったというか。僕やったら、家が火事になったら「焼き芋しよ!」ってなる。「芋ないんか!」って(笑)。火を消すよりも、燃えてるんやったら芋でも焼いとこかな、みたいな。

**CM: 朝起きて1番にすることは?**

KK: エア歯磨き。

CM: What kind of child were you growing up?

KK: In elementary and junior high school, I kept to myself, and I used to listen to the chatters of the popular kids in class and think to myself, "No, no, no that's not what you should say there," but it was all in my head, and I never actually joined the conversation or made people laugh. When I was in high school, one of my seniors who was a good friend of mine from the baseball team asked me to read over his script for a comedy show for the school festival He was the only one who listened to my quiet jokes and tell me that I was funny and he asked me for my advise. As soon as I saw the script, I said, "this is not funnty at all" (laughs). So I explained to him the nuances of what he should perform in the school festival, and he told me that I should participate instead That's when people around me started to realise how funny I was. Ever since then, people tried to get me into their groups, but I didn't want to stand out. much preferred to fantasize about things alone.

CM: Why did you move your base to Tokyo?

KK: Normally I would say it's because there are more microphones on stages in Osaka, but I will answer properly, since you served me some tasty bread for catering. It was simply because I thought there were more places to perform my skits and more people to perform in front of. At YOSHIMOTO, there is this unwritten rule to either get to the finals or win the award race before moving to Tokyo, but I didn't want to do that. If I was going to go, I wanted it to be at the weirdest timing possible. That's why I get a lot of people saying, "Oh, I didn' realise you were in Tokyo."

CM: When did you decide to pursue your career in comedy?

KK: I happen to be a comedian now, but it doesn't mean this is my goal. The reason I didn't start comedy right after graduating from high school was because I wanted to experience working first. I felt that once I start my career in comedy, I would never have an experience of working in the society, so thought, "It's now or never, I might as well try working now". I wanted to see how far I could go and how much I could be accepted into a workforce, so started out as a carpenter and tried various jobs such as working in a major construction company, an eel factory, a market, and a wedding hall. At the eel factory and the market, I was offered a full-time, and even got asked to be the next president of the company. I'm the kind of person that would be satisfied by that acceptance and quit the very next day. Whether it's drawing or being a comedian, I think it's okay to quit once I reach my level of satisfaction. As for what I'll do next, I don't know. I'm just doing these things to get to know myself My favourite is comedy, though, so I do centre my life around it.

CM: Who do you think is the funniest person in the world?

KK: Every human being is funny if you think about it. When I think of someone lying down in a penthouse of a high-rise apartment, it makes me giggle (laughs) They're so high up and they can just fall asleep. I also like to imagine a see through airplane flying above me so I can watch people watch movies in the air And think about people wearing masks over their ears — ears aren't supposed to be used like that, yet it's kind of cute and interesting that we wear masks like it's a normal thing. Everything is so funny.

CM: Who do you think is the most unfunny?

KK: People smoking in non-smoking areas or people who are not kind to the Earth... also, people who easily judge things. Comedy is usually evaluated simply by whether something is funny or not funny, but that is like seeing a painting in a museum and judging the art by whether it's well-drawn or not. think there's more to it, but these kinds of people who judge easily will probably walk out of a museum in 5 minutes, and I don't think that's funny.

CM: Who do you think is the most clever?

KK: Thomas Edison.

CM: Who do you think is the most stupid?

KK: Also Thomas Edison. I read that when his factory burned down, he called his son and said, "Go get your mother and all her friends. They'll never see a fire like this again." Isn't that amazing? He said, "I'll start all over again tomorrow." He was so smart, but such an idiot at the same time. I think he was just a person with all of his chakras wide open, a grown man with a child mind. If my house was on fire, I would say, "Let's roast a sweet potato! Does anyone have a sweet potato?" (laughs). Rather than putting out the fire.

CM: What's the first thing you do in the morning?

KK: Air-brush my teeth.

KK: 飼ってる猫と遊んでるときですかね。野良猫を3匹飼ってたんですけど、2匹はもう死んじゃって。今1匹残ってる子も、もう17年目くらいです。名前は大久保くん。特に由来はないです、降りてきたって言うか、右脳で考えました。

**CM: 今もっとも興味があるのは?**
KK: マッチョ。マッチョな方って、バーベル100キロとか上げると思うんですけど、あんな重たいものを持ち上げられるのに、寝るときは、この薄いまぶた1枚が上げられないんだと思ったら……。面白いし、マッチョもやっぱり人間なんだなと。小さいときから、そんなことばっかり考えてました。

CM: 好きなミュージシャンは?
KK: 僕も1曲出してるんですけどね……。『ポテンヒット』って歌。売れてないですけど。ミュージシャン仲間はいっぱいいますが、踊ってばかりの国、っていうバンドの歌はよく聴いてますね。と言いたいところをグッと堪えさせてもらいます。

**CM: 好きな映画は?**
KK: ロイ アンダーソン監督の『さよなら、人類』(2014年公開 / スウェーデン)っていうすごいシュールな映画があるんですけど、セットも街も、全部監督が自分で作ってるんです。役者もプロを使わずに、4年くらいかけて素人を集めて。むちゃくちゃ面白いですよ。

**CM: 好きな色は?**
KK: ゴールド。この前くっきー! さんに、ルミネの出番の合間に「靴買いに行こうや」って誘われて。そのとき僕は金の靴履いてたんですけど、「これよりも金の靴あったら買います」と。まさかないやろうと思ってたら、あったんですよね。で、それ買って。「あるんかい!」ってなりましたけど(笑)。真っ金金でした、ホンマに。金の靴は3足くらいですけど、他にも金のものは家にいっぱいあります。みなさんが今想像しているそのちょうど倍ぐらいあります。

CM: 地球上で1番嫌いなことは?
KK: 慣れ、ですかね。好きじゃないことをずっとやってたら、人ってやっぱり慣れてくるんです。慣れてくるが故に、自分が今イヤなことをしてるっていう意識が薄くなってきて、好きなことをしてる人を否定し出す。僕も我慢してもし早い段階でテレビに出て仕事をこなしてたら……。今頃、自分の好きなことをやっていないのにテレビ最高! みたいになってたかもと思うと、恐ろしいですね。

**CM: 今1番恐れていることは?**
KK: 携帯の設定のアップデートです。怖くないですか? 1回電源落ちて、だいぶん待つでしょ。あの間が1番怖い。大丈夫なんか? って不安しかないです。

**CM: 今、1番行ってみたい場所は?**
KK: 宇宙の外。宇宙って答える人は多いんでしょうけど、僕はそれより、宇宙の外に行ってみたいです。宇宙の外か、原宿。原宿に、エシカルなカレー屋さんがあるんですよ、化学調味料を使ってないような。そこか、宇宙の外。

**CM: 生まれ変わったら何になりたい?**
KK: 生まれ変わっても自分……ですかね。それか、女性パイロット。

**CM: 自分を一言で表すとしたら?**
KK: 愛、じゃないですか。

**CM: それは本気?**
KK: いや、早く終わらせたいな、と……(笑)。

**CM: 10年後、どうなっていたい?**
KK: と言いますと!?

**CM: これからチャレンジしたいことは?**
KK: 裸になって、仰向けで、スイカの種を飛ばして、へそに入れたい。それが百発百中で

KK: Probably when I'm playing with my cats. I used to have three stray cats, but two of them have already died. The one cat I have left now is about 17 years old. His name is Okubo-kun. I don't have a reason for his name, it just came to me from my right brain.

CM: What are you most interested in now?
KK: Macho people. They lift barbells that are like 100 kilograms, but when they sleep, they can't even lift their thin eyelids. I think that's funny, and it makes me realise they're also humans. These are the kinds of things I've been thinking about ever since I was little.

CM: Who are your favourite musicians?
KK: I've made a CD too… It's called "POTEN HIT" and it wasn't so successful. I do have a lot of musician friends. I would love to say that I listen to the band called "Odotte Bakari no Kuni" a lot, but I'm going to hold back.

CM: What is your favourite movie?
KK: There's a very surreal film called "A Pigeon Sat on a Branch Reflecting on Existence" (2014, Sweden) directed by Roy Andersson, where the director made the whole set and town by himself. He didn't use professional actors and gathered amateurs over a period of about four years. It's really interesting.

CM: What's your favourite colour?
KK: Gold. The other day, Cookie! asked me to go shoe shopping during the downtime between LUMINE show appearances. I was wearing gold shoes at the time, and he said, "I'll buy myself gold shoes if we can find any shoes more gold than those." I didn't think we would find anything, but we did, and I bought them. I was like, "Seriously?" (laughs). They were literally really golden. I have about three pairs of gold shoes, and I have many other gold objects in my house. Probably twice the amount of what you just imagined.

CM: What's your least favourite thing on earth?
KK: Getting used to things. If you continue doing something you don't like for a long time, you start to get used to it. Then you become less aware that you're doing something you don't like and start to judge people who are doing something they love. It's scary to think that if I were to have been on TV and had become successful early on, I could have been saying things like, "TV is the best!" by now, even if I don't enjoy what I would be doing.

CM: What are you the most afraid of now?
KK: Cell phone updates. Isn't it scary waiting for the phone to update? Like, is everything going to be okay? There's nothing but fear.

CM: Where would you like to go the most right now?
KK: Outside the universe. I'm sure many people would say space, but I'd rather go outside of space. Or Harajuku. There's an ethical curry shop in Harajuku that doesn't use chemical seasonings. That or outside the universe.

CM: If you were reincarnated, what would you like to be?
KK: If I were reincarnated, I'd become myself again… or a female pilot.

CM: If you had to describe yourself in one word, what would it be?
KK: Love, I guess.

CM: Really?
KK: No, I just wanted to get this interview over with (laughs).

CM: Where do you see yourself in 10 years?
KK: …and you mean!?

CM: What challenges would you like to take on?
KK: I want to get naked and lay on my back and fly a watermelon seed into my belly button. I'd like to get it right and never miss a shot. I'd also like to make videos and movies. Acting is interesting too, but I have more interest in directing.

CM: What is your motto?
KK: If you have time to challenge your limit, make pasta instead.

jacket & shirt LITTLEBIG
nasubi & sunglasses MODEL'S OWN

# I'M A SUPER LUCKY BOY

## interview with KENJI YAMAUCHI

photos_Yume Ippei fashion_Nao Koyabu grooming_Shuco @3RD
model_Kenji Yamauchi from KAMAITACHI @YOSHIMOTO
fashion assistant_Mayuko Saito
grooming assistant_Takako Koizumi

DO YOU WANT TO SEE MORE ?

top **KIDILL @WALL HARAJUKU**  hat **HIZUME** 

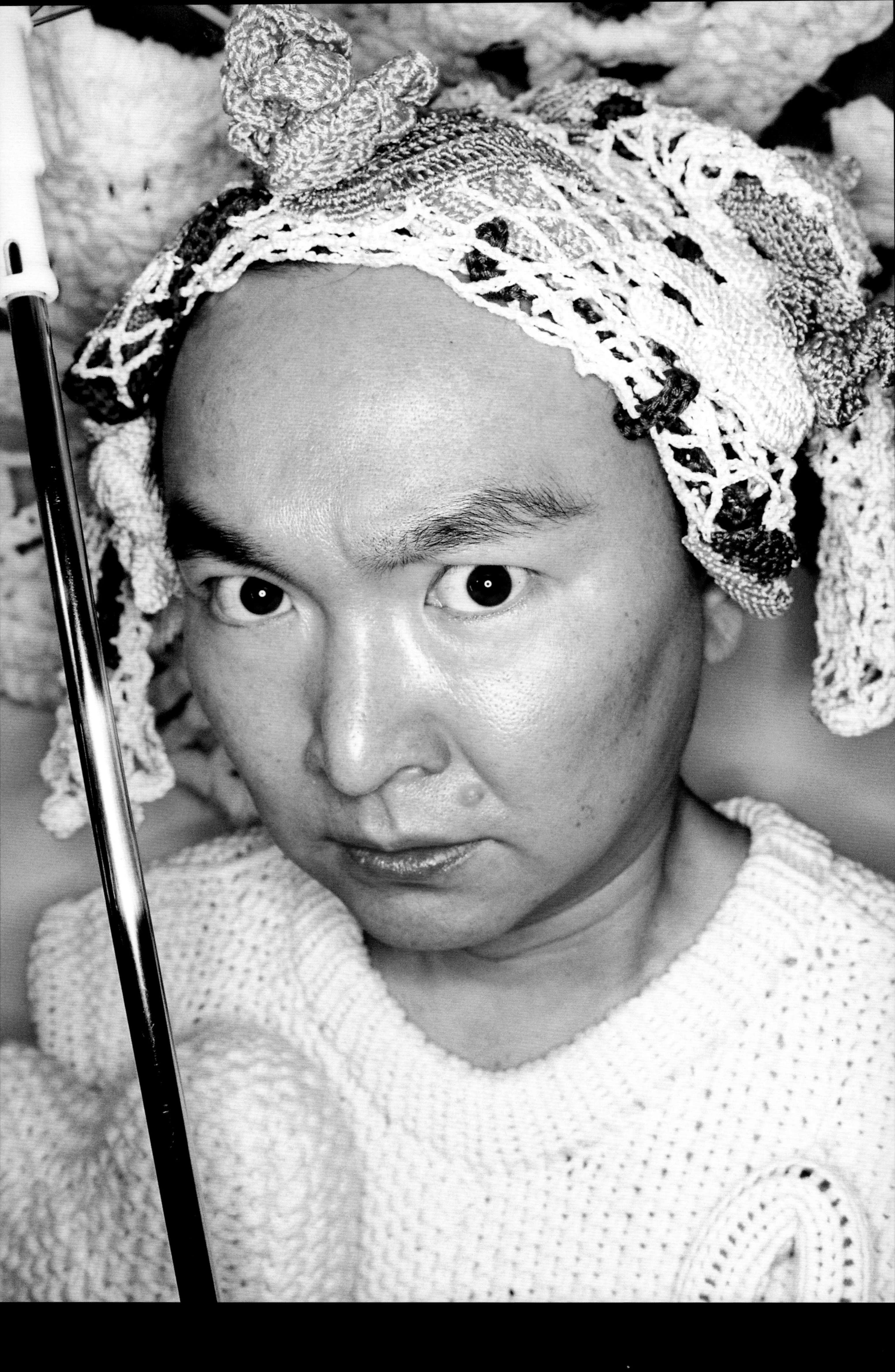

jumpsuit, crochet worn as headpiece & umbrella **PILLINGS**

commons&sense man (以下CM): お笑いの世界に入ったきっかけは?
Kenji Yamauchi (以下KY): 小さい頃からお笑いの世界に入ろう、芸人っていいなぁと思ってたんですけど、最終的に学校の先生になるか芸人になるかでめっちゃ迷ってました。でも教育実習のときにえらいスベって、生徒に受け入れられなかったのをきっかけに、芸人の道に進もうと決めました。教育実習でウケてて生徒の反応が良かったら、教員になってたかもしれません。

CM: 具体的に、どうスベった?
KY: 1番スベったのは、体育館に全校生徒を集めて、今後の進路についての質問に答えるっていう会合をやったとき。「社会人になったり大学生になったり、いろんな道があると思いますけど、僕はこの松江東高校(編集部注:山内さんの母校)が1番好きです!」って言った後、自分の高校時代の体操服をそのとき着てた服の下に仕込んでたんで、ズボンをバンっておろしたら体操服を着てる、っていうのをやって、ホンマにスベって……。無言で履きなおしました。体操服は上下仕込んでたんですけど、下だけ見せて、やめました……。ジャージの長ズボンやったんですけど。

CM: 目指している"お笑い"は?
KY: 芸人仲間からも面白いと言ってもらえるお笑い、ですかね。お客さんは笑ってくれてても、芸人に「いや、そんなんで笑いとっても……」とか思われてるのはイヤですから。

CM: お笑いにおいて必要不可欠なこととは?
KY: 情報と学力。確かビートたけしさんがおっしゃってたと思うんですけど、最近はかなりの数の人が大学を出て社会人になってるから、自分がその人たちよりもモノを知らなかったら笑わせることなんかできない、だから芸人を目指しているなら大学出てからでもいいぞ、みたいなことを昔に聞いた記憶があって。ホンマにそうやなと思って、その言葉にめっちゃ影響受けました。

CM: "お笑いの神が降りてきた"と思った瞬間は?
KY: ないんですよね、全部自力です。まだ神の力は借りてないです。

CM: インスピレーションの源は?
KY: 日常で面白いなと思ったちょっとしたことをメモしておいて、ネタを作るときにそこから広げていく感じです。普段生活する中で、何か変やなと思ったことが、結構ネタの源になってます。たとえばトトロのネタ(ジブリ映画『となりのトトロ』を山内さんは見たことがない、ということを自慢話として披露する漫才)は、濱家と普通に喋ってるときに「俺、トトロ見たことないねん」って言ったら、「それでちょっとマウント取ろうとしてる感じ、なんやねん」みたいになって。で、これはネタでも行けるんちゃうか、と。

CM: ネタはどうやって膨らませる?
KY: 最初は1人で考えて、ある程度「こんなん面白いと思うねんけど」っていうところまで固めてから濱家に話して、それを濱家も面白いと思ったらネタにしていきます。濱家の反応が悪い場合は、僕だけが面白いと思ってる可能性が高いんで、ネタには至らないですね。独りよがりのネタにならないように、気は遣ってるつもりです。

CM: その点で、濱家さんのセンスを信頼している?
KY: そうですね。自分としては普通のお客さんにもわかるやろうと思ってることでも、濱家が「いや、それはあんまり伝わらへんのちゃうか」という場合、大抵お客さんがホンマにわかってなくて、濱家が正しいことが多い。だから、万人に伝わる、伝わらないのラインは、大体濱家に確認します。

CM: その濱家さんが「イってる」と表現する狂気の眼付きになった、いわゆる"無双状態"にあるときの感覚は?
KY: これは仕事、という気持ちのみです。

CM: ここ最近で1番笑ったのは?
KY: 後輩のビスケットブラザーズのネタ。

CM: 今の日本のお笑いをどう思う?
KY: 売れるべき人が売れていると思います。

**commons&sense man (hereafter CM) : What got you into comedy?**
Kenji Yamauchi (KY): I've wanted to become a comedian ever since I was a little kid, but I was really torn between that or becoming a teacher. When I bombed a joke at my educational training, I decided to become a comedian instead. If I had been funny and the students had responded well to my joke, I might have become a teacher instead.

**CM: What exactly was the joke?**
KY: When the students were gathered in the gymnasium to answer questions about their future careers, I secretly wore the school's uniform gym clothes under my outfit and said, "I know there are many paths you can take, such as working or becoming a university student, but I love Matsue Higashi High School the most!" while pulling down my pants to reveal the gym clothes. It went so badly that I quietly put my pants back on without a word. I didn't even have a chance to reveal the top.

**CM: What kind of comedy do you strive for?**
KY: I want to do comedy that fellow comedians will find funny. I wouldn't want to do jokes that only the audience responds to and comedians will say isn't funny.

**CM: What is essential in comedy?**
KY: Knowledge and education. Beat Takeshi (Takeshi Kitano) has said this before, but with the number of educated people out there these days with university diplomas, without at least the same amount of knowledge or more, you wouldn't be able to make them laugh. He said that if you want to be a comedian, you can do it after college, and that really stuck with me.

**CM: Was there ever a moment when you felt the power of a comedy God?**
KY: No, everything I've done has been on my own without help from a God.

**CM: Where do you get your inspiration?**
KY: I write down little things that I find funny in my daily life, and then I expand from there when I'm creating a skit. For example, my Totoro skit (a skit in which Yamauchi brags that he has never seen the Ghibli film "My Neighbor Totoro") came about when I was talking with Hamaie and I said, "I've never watched Totoro," and he said, "Do you think you're better than me because you've never seen Totoro?" And I thought to myself, "Oh, this could work as a skit."

**CM: What's your process of creating skits?**
KY: I first think of skits alone and after I solidify an idea to a certain extent, I tell it to Hamaie. If he agrees that it's funny, we turn it into a skit. If he doesn't respond to it well, there's a good chance that I'm the only one who thinks it's funny, so I scrap it. I try to be mindful not to create self-indulgent skits.

**CM: It seems that you trust Hamaie's opinion in that sense.**
KY: Definitely. Even if I think something would be obvious to the audience, if Hamaie says, "No, nobody would get that," he's usually right and the audiences won't get it. That's why I usually check with him on what could be easily understood by the crowd.

**CM: As Hamaie puts is, you often get into the state of trance with your "crazy look" in your eyes. How do you actually feel when you're in that state?**
KY: I remind myself that it's my job.

**CM: What's the funniest thing you've seen recently?**
KY: A skit by BISCUIT BROTHERS, the younger generation comedian duo.

**CM: What do you think of the current comedy scene in Japan?**
KY: People who deserve to be successful are succeeding.

**CM: What are you into at the moment?**
KY: Sneakers. They bring me luck, helps me get jobs, and I think they bring me a to a higher stage.

**CM: 今、特にハマっているものは?**
KY: スニーカー。運気が上がるし、仕事にもつながるし、自分をより上のステージに連れてってくれるんで。

**CM: 今もっとも興味があることは?**
KY: やっぱり子どもの成長ですね。日々感じてるんですけど、子どもがなんとなく面白いことを言ってるような、それがボケっぽいことのような気がするんです(笑)。でも僕、家でふざけたりすることはほとんどないんで、どこでこんな風にチョケる"気配"を身につけたんかなって、ちょっと不思議ですけど。

**CM: その様子がいずれネタに発展する可能性も?**
KY: そうですね。ただ、家族やから面白いんかなっていう感じもあるんで、ただのノロケになりそうですけどね(笑)。

**CM: 尊敬する人は?**
KY: ダウンタウンの松本さん。1番影響を受けてるんで。当時から面白くて、今でもキレッキレでずっと面白いのは、やっぱりすごいなって思います。

**CM: 2019年のM-1グランプリでは、松本さんだけが、決勝でかまいたちに票を入れていましたね。**
KY: いや、ホンマに嬉しかったですね。あれがあるとないとでは、全然違ってたと思います。

**CM: この世で1番おもしろいと思う人は?**
KY: (千鳥の)ノブさんかなぁ。普段から、あんな感じなんですよ。で、たまに見せるポンコツな感じとか、返しの鋭さとか……。めちゃくちゃ面白いなと思います。舞台とか収録中だけじゃなくて、普段も面白い人なんです。

**CM: 1番おもしろくない! と思う人は?**
KY: クロスバー直撃っていうコンビの渡邊さん。先輩ですけど。普段からポンコツで、みんなから愛されてるのに、『アメトーーク!』とかのでっかい仕事が入ったとき、"全ポンコツ"なんです。大事なときに「お、ハネたな!」とならない。普段はめっちゃハネるのに、なんでやねん、と。ちょっと"もったいな"って思う部分はあります。

**CM: 1番"賢いな!"と思う人は?**
KY: ネタで言うと、和牛の水田かな。構成力とか、ネタとして取り上げてる(他人や物事の)イヤな部分を見てると、"ひねくれ賢い"なぁって思います、めっちゃ。

**CM: 1番"アホやな〜"と思う人は?**
KY: クロスバー直撃の渡邊さん(笑)。ホンマに。ホンマにダメなんですよ、めっちゃアホやと思います。大事なとこで全部ダメにする。それを含めて、先輩ながら可愛いなと思ってるんですけど、なんかそれは伝わってないみたいです。

**CM: 地球上で1番嫌いなモノは?**
KY: 虫。もう虫全般がダメで。たとえば、蛾はもちろんですけど、蝶もダメ。カブトムシもダメですし。虫を採取するロケとか、ホンマに無理です。島根(編集部注:山内さんの出身地)にも虫はいっぱいいましたけど、気持ち悪くて。芸人として大概のことはOKなんですけど、夜の虫採取ロケだけはマジで無理ですね。

**CM: あなたが今1番恐れているコトは?**
KY: スキャンダルです。出さないように努力はしてますけど、最近はもう、何年前の話が出てくるかもわからないんで……。もし週刊文春の記者が突撃してきたら、僕は本気で走って逃げます。

**CM: 生まれ変わったら何になりたいですか?**
KY: う〜ん……。また自分、でいいかな。今、何の不自由もなく生きてるんで。

**CM: What are you most interested in now?**
KY: Watching my children grow up. I always feel like my kids are natural-born comedians without trying to be (laughs). I rarely joke around at home, so I wonder where they get their sense of humour.

**CM: Do you think that will eventually develop into a skit, too?**
KY: It's a possibility. But I think it's only funny to me because it's my family, other people would probably think I'm bragging (laughs).

**CM: Who do you look up to?**
KY: Hitoshi Matsumoto from DOWNTOWN. He's influenced me the most. I think it's amazing that he's been funny since the beginning and still continues to be sharp and funny all the time.

**CM: During the 2019 M-1 Grand Prix, we saw that he was the only one who voted for KAMAITACHI at the finals.**
KY: That made me so happy. Without that vote, the outcome would have been totally different.

**CM: Who do you think is the funniest person in the world?**
KY: I think Nobu from CHIDORI. He's the same in real life as he's on TV. His balance of stupidness and sharpness is super funny to me, with or without being on a show.

**CM: Who do you think is the most unfunny?**
KY: He's older than me, but Watanabe from the duo called CROSS-BAR CHOKUGEKI. He's usually dopey and everyone loves him, but when he got to be on a big show like "Ametalk!", he was dopey the whole damn time. At key points, he'd usually hit a few good laughs, but for some reason, he couldn't. He's such a waste of talent, sometimes.

**CM: Who do you think is the most clever?**
KY: In terms of skits, I'd say Mizuta from WAGYU. When I see his ability to compose unpleasant parts of things and people he sees into skits, I really think he's a twisted and clever person.

**CM: Who do you think is the most stupid?**
KY: Again, Watanabe from CROSS-BAR CHOKUGEKI (laughs). Seriously, he's really that bad and stupid. He ruins everything at the most important part. All in all, I think it's cute, but I don't think he gets it.

**CM: What do you hate the most?**
KY: Bugs. I hate all bugs. Moths and even butterflies. Beetles, no way. Shows, where they ask me to collect bugs, are my worst nightmare. Shimane where I'm from has lots of bugs and they were disgusting. As a comedian, I'm okay with most things, but nothing would be worse than having to collect bugs at night for a TV show.

**CM: What are you the most afraid of now?**
KY: Scandals. I try my best to avoid it at all cost, but these days, you never know how many years of dirt people are willing to dig up. If a reporter from the Weekly Bunshun came at me, I would run for my life.

**CM: If you were reincarnated, what would you like to be?**
KY: Hmmm... I'd be fine to become myself again. I have no complaints about my current life.

**CM: また同じように芸人になって、同じ相方と組む？**

KY: あ、もう濱家は結構です。自分で何とか……（笑）。そもそも芸人をやるかどうかもわからないですけど、なんとなく、今のこの自分の感覚は持ったままで生まれ変わって、今度はいろいろ違うことをやってみたいなっていう気持ちはあります。僕、実はいろんな旅をしたかったんですけど、この仕事してたら、1年休んで旅行行くとかはできなさそうですよね。この間、『YOUは何しに日本へ？』（テレビ東京）を見てたら、普通に会社に勤めてて、定年を迎えてから2人で世界を周ってるっていう夫婦が取り上げられてました。僕もそんな風に、芸能界を引退して、60歳くらいから世界を旅するのはアリかなと思いました。もう、“NOボケ”で。一切ボケたりせずに、ただただ普通に歩くだけ（笑）。

**CM: 自分を一言で表すとしたら？**

KY: スーパーラッキーボーイ。今改めて思い返してみると、タイミングとか、結果的に全部がうまくいってるなと。M-1では、実はコンビ結成して3～4年目くらいで1回準決勝行ってて。僕が中国人に扮するネタで、決勝まで行ってもおかしくないくらい、めっちゃウケたんですけど、結局決勝には行けなかった。そのときは「なんでやねん！」って思ってましたけど、もしあのときにそのネタで決勝に行ってたら、そのあともずっと中国人キャラを求められて、実力もないのにテレビに出てそのキャラで続けなアカンっていう状態になってたやろうなと。そう思うと、あそこで落とされてよかったんですよ。結果、すべてがいい方に転んでるなと思います。YouTubeをはじめるタイミングもそう。他のみんなは、ほとんどコロナ禍をきっかけにはじめてますけど、僕らはコロナが広がりだすちょい前で、みんなよりちょっとだけ、1ヶ月半くらい早かったんです。それもラッキーでしたね。

**CM: YouTubeをはじめた理由は？**

KY: 2019年のM−1で“バトル”が終わったので、新たにバトルするために乗り込みました。

**CM: 10年後、どうなっていたい？**

KY: ゴールデンで、かまいたちと言ったらコレ、っていう番組が1本か2本あって、今みたいに毎日働かずに、レギュラー稼働はするけど1本のギャラがめっちゃ高い感じで、休みが多いっていう生活をしてたいですね。今それができてるのは、ダウンタウンさん、あと僕らはいつも言ってるんですけど、大阪のトミーズさんくらいですかね。トミーズさんは、ホンマに稼働が少ない。ダウンタウンさんとか（明石家）さんま師匠は、まだまだ「働きたい」っていうバトル精神がすごく強いと思うんですけど、トミーズさんは多分もうそれがないんです（笑）。舞台も出られてますけど、多すぎず、「自分らはこれくらいでいい」っていう量で。あとはちょうどいいくらいのテレビの稼働しかしてない。それで、大阪ではもう確固たる地位を築いてはりますから。その生き方はいいなと思いますね。

**CM: その拠点は東京を想定している？**

KY: いや、これは濱家も同じこと言ってるんですけど、何年か経って、自分らで納得できるというか、「もういいかな」と思えるとこまで行ったら、大阪に戻りたいなと。ただ、ホンマに「もういいかな」と思えるかどうかはわからないですけど。

**CM: 今後やってみたいことは？**

KY: 俳優業ですね。でも稽古嫌いなので、ちょい役で……。

**CM: 最後に、座右の銘を教えてください。**

KY: 「為せば成る　為さねば成らぬ何ごとも　成らぬは人の為さぬなりけり」っていう名言があって、それはホンマにその通りやなと。努力をしてなかったときは、基本的に結果もついてきてなくて、頑張ったときは、それなりの結果がついてきてるっていう気がするんで。何でも「出来る！」って思ったら出来るなぁと。

**CM: Would you like to become a comedian again, and work with the same partner?**

KY: Oh, I don't need Hamaie. It'll be fine with just me alone (laughs). I don't even know if I'd be a comedian, but if I were to live my life again, I'd like to travel more. With my current job, I can't just take a year off to travel. The other day, I was watching "Why Did You Come to Japan?" (TV Tokyo), and there was a couple who had normal careers and started travelling the world together after retirement. I thought it would be nice for me to do the same, retire from the entertainment industry and start travelling the world at the age of 60. No more joking around, just normal walking around (laughs).

**CM: If you had to describe yourself in one word, what would it be?**

KY: A super lucky boy. We actually made it to the semi-finals in the M-1 Grand Prix in our third or fourth year as a duo. We did a skit where I played a Chinese person, and it was so popular that it could have made it onto the finals, but in the end, we didn't make it. At the time, I thought, "Why didn't we win?", but if we had made it to the finals with that skit, I would have been asked to play a Chinese character for the rest of my career, and have no chance to grow. So when I think back at the time, I'm happy that we didn't win. Everything turned out for the better, including the timing of when we started our YouTube channel. Most people started because of the pandemic, but we started just a month and a half earlier than everyone else, so we got lucky.

**CM: What made you want to start a YouTube channel?**

KY: Our 2019 M-1 Grand Prix battle was over, so we wanted to start a new battle on our own.

**CM: Where do you see yourself in 10 years?**

KY: I'd like to have one or two primetime shows our duo is known for, and not work every day as we do now. I'd like to have regular shows that pay enough to have the rest of the week off. The only people who I know can do that now are DOWNTOWN and TOMMY'S in Osaka. TOMMY'S really don't do much. I think DOWNTOWN and Sanma (Akashiya) still have the drive to work, but TOMMY'S probably don't have that anymore (laughs). They still perform on stage, but just enough, and a little bit of TV time here and there, and they're legends in Osaka, so I respect that way of living.

**CM: By then, do you still want your home base to be Tokyo?**

KY: Hamaie and I talk about this all the time, but after a few years, when we are satisfied with our work, or when we feel like we have done enough, we would like to return to Osaka... although I don't know if we'll ever really be able to say we've done enough.

**CM: What challenges would you like to take on?**

KY: Acting. But I hate rehearsing, so just a small role for me is fine...

**CM: What is your motto?**

KY: There the saying that goes, "If you don't try, you will never succeed," and totally agree. If you don't try that hard, the results end up being so-so, and the more effort you put into something, I think the results follow, so I try to believe that I can do anything when I'm trying something.

jumpsuit, dress used as a bag & crochet worn as headpiece **PILLINGS**

# THE TOKYO MATCH 05

*CHIKASHI SUZUKI*

鈴木家之墓
榎本家之墓

ガンバレ
東日本

**鈴木 親** フォトグラファー

1972年生まれ。1998年渡仏。雑誌Purpleにて写真家としてのキャリアをスタート。国内外の雑誌から、ISSEY MIYAKE, TOGA, CEBIT, GUCCIのコマーシャルなどを手掛ける。
主なグループ展: 1998年 COLETTE（パリ / フランス）、2001年 MOCA（ロサンゼルス / アメリカ）、2001年 HENRY ART GALLERY（ワシントン / アメリカ）
主な個展: 2005年 TREESARESOSPECIAL（東京 / 日本）、2009年 G/P GALLERY（東京 / 日本）、2018年 KOSAKU KANECHIKA（東京 / 日本）
作品集: 2005年『shapes of blooming』(TREESARESOSPECIAL)、2008年『Driving with Rinko Kikuchi』(THE INTERNATIONAL)、2009年『CITE』(G/P GALLERY, TREESARESOSPECIAL)、2014年『SAKURA!』(LITTLE MORE)

**Chikashi Suzuki** **Photographer**

Born in 1972. Started his career as a photographer with Purple Magazine (France) in 1998. Currently shoots for international and domestic magazines, as well as commercial photos for ISSEY MIYAKE, TOGA, CEBIT, GUCCI etc.
Group exhibitions: COLETTE (Paris, France) in 1998, MOCA (L.A, USA) in 2001 and HENRY ART GALLERY (Washington, USA) in 2001.
Solo exhibitions: TREESARESOSPECIAL (Tokyo, Japan) in 2005, G/P GALLERY (Tokyo, Japan) in 2009 , and KOSAKU KANECHIKA (Tokyo, Japan) in 2018.
Has published following books: “Shapes of Blooming” (TREESARESOSPECIAL) in 2005, “Driving with Rinko Kikuchi” (THE INTERNATIONAL) in 2008, “CITE” (G/P GALLERY, TREESARESOSPECIAL) in 2009 and “Sakura!” (LITTLE MORE) in 2014.

# IT'S SUCH A MIRACLE

## interview with CAMPBELL from THE MISTRESS 5

illustrations & photos_The Mistress 5

近頃ちょくちょく見かけるようになった、モンスターのような5人組。その名はTHE MISTRESS 5。全身真っ黒で、ちょっぴり怖い感じもするけれど、よく見るとなんだか愛らしい。ちょっと前にはとあるイベントで、ついこの間はオシャレな百貨店のショーウィンドウでも目撃。InstagramやTwitter、YouTubeでも、ちょっぴり話題になりはじめている。どうやら彼女たちは、自然環境の大切さを訴え、人間だけではなく地球上に暮らすあらゆる生き物たちを大切にしながら、かつての美しい地球の姿をみんなで取り戻そうと活動しているらしい。しかも5人それぞれが異なる"問題"を受け持っていて、地球の人たちと手を取り合って、解決に向けて動いているんだって。宇宙の中でも飛び抜けて美しいと評判の地球に憧れて、宇宙からやってきたっていう噂だけど、一体全体宇宙のどこから、しかもどうして地球にやってきたんだろう？本当は何者？地球で何をしているの？そこで今回は、ベールに包まれた真の姿に迫るべく、そのなかの1人、地球温暖化問題担当のキャンベル（Campbell）にインタビューを敢行した。

Recently, we've been seeing a lot of THE MISTRESS 5, the monster-like group. They're all black from head to toe and are a bit scary looking, but if you take a closer look, they're kind of cute. We saw them at an event once, and recently in the display window of a fashionable department store just the other day. They're starting to become popular on Instagram, Twitter, and YouTube. Apparently, they are working to bring back the beauty on Earth by stressing the importance of the natural environment and taking care of not only humans but all living creatures on the Earth. Each of the five members is in charge of solving a different issue the planet faces, hand in hand with earthlings. Rumour has it that they came from another planet that admires the environment of the Earth, which has a reputation of being the most beautiful planet in the universe. Where in the universe did they come from? Who are they, and why did they come to visit? We had a chance to speak to Campbell, who is in charge of the prevention of global warming, to find out more.

**commons&sense man** (以下CM): **まず、基本的な質問から。おいくつなんですか?**
キャンベル(以下C): 地球の年齢で言ったら、139歳。

**CM: どこから来たんですか?**
C: 正確には言えない。でも、ウルトラマンとかスーパーサイヤ人とか、地球にも宇宙からきた人はたくさんいるでしょ? そのなかでも私たちの星が1番遠いと思うわ。

**CM: どんな星なのでしょう?**
C: とてもいい星よ。でも地球ほど恵まれてはいないと思う。この星は、海や空が青くて緑があって、本当に奇跡だと思うわけ。地球のみなさんは当たり前と思ってるかもしれないけど、こんなに環境に恵まれた素敵な星は宇宙でも他にはないもの。それを忘れないでほしいな。

**CM: 地球のこと、どうやって知ったんですか?**
C: テレビで見たり、学校で習ったりしたし、私たちの星でも憧れてる人はいっぱいいる。でも、その反面…… なんて言うか、怖い星でもあるって聞いてたけど。

**CM: それはどういう怖さでしょうか?**
C: 人は700万年くらい前に地球に生まれてからずーっと、戦争とか紛争、色んな戦いを延々と続けているでしょう? 今、この瞬間も、世界のどこかで誰かが戦ってる。ずーっと戦争してるなんていう星、他に聞いたこともないもの。

**CM: キャンベルさんの星では戦争はなかったんですか?**
C: ないわ、まったく。愚かなことだもの、同じ仲間同士で殺し合うなんて。

**CM: それでも地球に来たのは?**
C: やっぱり自然環境とか、地球のみなさんが作ったイノベーションが素晴らしいって聞いてたから、死ぬまでに1度は見ておきたいなって。それで、卒業旅行で。仲良しの5人で集まって、行きましょうって。

**CM: 初めて地球に降り立ったときの感想を教えてください。**
C: それはそれは感動したわ! 最初はカナダだったっけ、東の方のケベック。メイプルシロップが採れるでしょ? 大好きなの。イミテーションしか食べたことがなかったから、まずメイプルシロップをいただいたわ。でも、実を言うと……。メイプルの木に吸い付いて、直接いただいちゃったわけ(笑)。だから、怒られたら困ると思って公表はしてなかったんだけど。地球では、ハチミツの方が重宝されてるんでしょう? 蜂さんが一生懸命頑張って、命をかけて貯めた蜜を根こそぎ持っていくなんて、一体どういうこと? って思っちゃうけど。

**CM: キャンベルちゃんは女の子?**
C: そうよ。でもそれもあなたたちの考え方で、性別っていう概念、あんまりないもの。

**commons&sense man** (hereafter CM)**: Let's start from the basics. How old are you?**
Campbell (hereafter C): In Earth years, I'm 139.

**CM: Which planet are you from?**
C: I can't tell you exactly where, but you know many people that came from outer space like Ultraman and Super Saiyan, right? The planet we come from is probably the farthest from the Earth than any of them.

**CM: What kind of planet is it?**
C: It's a very nice planet, although it's not as blessed as you are here. It's practically a miracle that the Earth has blue oceans, blue skies, and greenery. You Earthlings take it for granted, but please don't forget that there's no other planet in the universe with such a wonderful environment.

**CM: How did you learn about our planet Earth?**
C: I've seen it on TV, learned about it in school, and there are many people on our planet who want to live on Earth. But I've also heard that it's kind of a scary planet, too.

**CM: What kind of scary?**
C: Since people were born on Earth about seven million years ago, you have been fighting wars against each other as well as many other conflicts and battles. Even as we speak, someone is fighting someone, somewhere in the world. I've never heard of any other planet that's been at war for so long.

**CM: Are there wars on your planet?**
C: No, not at all. It's stupid to fight and kill your own people.

**CM: What made you still want to visit our planet?**
C: I've heard that the natural environment and Earthling-made innovations on Earth are amazing, so I wanted to see it at least once in my lifetime. This is my graduation trip. I'm visiting with four of my best friends.

**CM: How did you feel when you first landed on Earth?**
C: It was quite an experience! The first place we landed was Canada, I think, in Quebec. Do you know that they make maple syrup? I love it. I've only had imitation maple syrup, so it was my first time. To tell you the truth, I directly sucked maple syrup out of the tree myself (laughs). I never told anyone because I didn't want to get in trouble. Honey is more valuable here, right? I wonder what the point is of taking away honey that the bees spent their whole lives accumulating.

**CM: Are you a girl?**
C: Yes, I am, in your context of gender, because we don't really have genders.

**CM: これまで地球をいろいろ探索してきた中で、1番印象的だったのは？**
C: 私たちの故郷に近いなと思ったのは、アイスランド。何にもないけど、一部に人が住んでるみたいな。星ってどこもそういう感じだけどね。

**CM: 変身されたことのお話をちょっと詳しく聞かせてもらえますか？**
C: 憧れの星だから、すごく期待してたのに……。オーストラリアに行ったらなんかずっと火の匂いがしてるし、アメリカのマイアミでは満潮時になったら街が水に浸かっちゃうし、なんか思ってたのと違うなって。海岸沿いに行ったときも、観光地の海は綺麗だけど、ちょっと外れたらもうゴミだらけ。北極とか南極に行っても氷は溶けちゃって大変なことになってた。これ、私たちみたいな第三者からしたら、崩壊してるんじゃないかって。環境破壊とかいう賢そうな言葉を使ってるからアカデミックに聞こえるけれど。空気も思ったより綺麗じゃないし、なんかそういうところばかり見てたら、悲しくなって悲しくなって、どんどんそれが怒りに変わっていって……。

**CM: THE MISTRESS 5 それぞれのミッションを教えてください。**
C: 私に関しては、地球温暖化。ターリントンは、道徳問題。人の差別もそうだけど、これってすごく根深い問題なのよ。人の陰で辛い思いをして死んでいく生き物たちもなんとかしてあげなきゃ。で、クロフォードがゴミ問題。地球温暖化とも直結してるから、2人で連携していくことも多くなるかな。クリステンセンは水問題。飲むための水を確保するためにダムは必要、でもダムは自然破壊に繋がるし……。矛盾だらけの中でかろうじて保ってたバランスが、いよいよ崩れはじめてるの。そしてエヴァンジェリスタが海洋森林問題。地球温暖化、ゴミ問題にも直結してる大きな問題よ。

**CM: 実際にどんな活動をしているんですか？**
C: 日本のアパレル会社の社長が、安いダウンを作って、それをまたリサイクルするとか言ってたけど、ちょっとクエスチョンよね。知り合いのシベリアのネネツ族の人たちは、ずっと同じ洋服を着てるわ。子ども時代に作った銀狐の靴を、一生履き続けるのよ。零下30度のアラスカで、夕食の1時間だけ火を焚いて暖かさを楽しんで、終わったら消して、また零下30度。今、皆さんサスティナブルって言うけれど、これこそ究極のサスティナブル。リサイクルどうこうっていうよりも、まずゴミも、二酸化炭素も出さないことが1番大切だと思う。今のところ、ポイ捨て防止のために10枚入りのリサイクルできるゴミ袋を作ったり、ステッカーを作ったり。ステッカーは、スマートフォンに貼ってもらったりして、見るたびに「電気消さなきゃ！」とか「このコンセント、抜いておこう」「エアコンの温度設定、ちょっとみてみよう」とか、身近でできることを思い出してもらえたらいいな。あとは、SDGsもうわべのトレンドで終わらないように、啓蒙活動みたいなことはちょっとしていかなきゃって。一般企業さんのイベントはもちろん、国連関係とかNPO、NGOみたいな団体のイベントにも積極的に参加していくつもり。そこでみなさんと実際にコミュニケーションをとって、一緒に盛り上げていきたいな。あとは、地球にはYouTubeとかInstagramとかがあるでしょう？そういうところでもいろいろ配信していくから、ぜひ見てみてね。最終的にはみなさんにアクションを起こしてもらって、地球温暖化の数値、すごく良くなってきたよ、天候も落ち着いて台風も少なくなったよ、とか、街が水没しなくなったよ、みたいになったら最高。

**CM: Of all the places you've explored on Earth, what made an impression the most?**
C: Iceland felt the most like my home planet. There's nothing there, with just a few inhabitants. I guess it's how most planets are.

**CM: Can you tell us about your transformation?**
C: Earth is a planet I've always wanted to see, so I was really excited about visiting. But when I went to Australia, it smelled like fire all the time, and in Miami, the city became flooded during the high tides. When I visited the coasts, the tourist spots were beautiful, but nearby were places with garbage everywhere. Even when I visited the North and South Poles, the ice had melted and it was a mess. It seems like the Earth is breaking down. People here use terms like environmental destruction, which makes it sound more difficult than it actually is. The air is not as clean as I thought, either, and all of these things made me sadder and sadder until it turned to anger.

**CM: What are the missions of each of THE MISTRESS 5 members?**
C: My mission is to fight global warming. Turlington's mission is to tackle moral issues protecting the vulnerables, which is a deep-rooted problem. We also have to do something about the creatures who die because of humans. Crawford is working on the waste crisis, which relates to global warming, so I think the two of us will be working together a lot. Christensen is in charge of the water issues. Dams are necessary to secure drinking water, but in the process of their creation, it destructs nature. All of these contradictions that have barely maintained their balance are now beginning to collapse, and it's showing up in the environment. Evangelista is tackling marine forest issues, which is a big problem adjacent to global warming and the waste crisis.

**CM: What do your activities involve?**
C: We heard the president of a Japanese apparel company said he wants to make cheap down jackets and then recycle them, but that kind of mindset is questionable to me. The Nenets tribe people in Siberia I've met wear the same clothes for life. They make silver fox shoes as children and wear them for the rest of their lives. In Alaska, where the temperature is negative 30 degrees celsius, they build a fire for just an hour for dinner, enjoy the warmth, then turn it off, and go back to living at negative 30 degrees. People talk about sustainability, but this is the ultimate form of sustainability. Whether or not you recycle is not the main issue, but to stop the emission of carbon dioxide. So far, we've made 10-pack recyclable garbage bags and stickers to try to prevent littering, as well as smartphone stickers to remind people to turn out the lights, unplug their outlets or lower their air conditioner setting. We need to educate people about the SDGs so that they don't end up being a mere superficial trend. We plan to actively participate in events held by companies, as well as those held by organisations including the United Nations, NPOs, and NGOs. We'd like to communicate with people at those events and make things happen. The Earth has YouTube and Instagram, and we'll be making announcements on those platforms as well, so please check them out. My ultimate goal is for everyone to take action, and slowing down the speed of global warming, making the weather more controlled with fewer typhoons, and have fewer cities be flooded.

CM: 今、周りの状況を見て思うことはありますか?

C: SDGsって、なんか人を中心に考えてる感じがする。弱者保護にしてもそうで、今の世の中には、ハラスメントっていうのがいっぱいあるでしょう? あなたたちの日本でも、なんでもそういうハラスメントみたいになっちゃう。でもそれを言い出したらペットだってそうだし、動物園とか水族館とか、植物園も含めて、全部ハラスメントだって思わない? 教育だ何だって偉そうな理念を掲げてるけど、結局全部人のエゴ。ペンギンさんが見たかったら、お金を使って、頑張って南極とか北極まで行って見てください、ね。

CM: それでも人のことが嫌いになったりはしませんか?

C: もちろんしないわ。欲望があるからこそ、色々発展するんだし。いろいろな生き物の中で、唯一人だけが相手の立場になって物ごとを考えられるし、優しくもできるんだから。だから私は諦めてないし、みんなが優しくなれば、もうちょっと世の中が楽しくなるんじゃない?争いをしても、悲しみと憎しみしか残らないもの。

CM: 悲しみと憎しみがない星は素晴らしいですね。

C: 私たちはその悲しみが怒りに変わって、こんな姿になっちゃったけど……。いま日本では、仲良しのCUBEっていう会社がやってるcommons&senseって雑誌(この雑誌)のチームと、イベントとかPR活動をしているNIKOLA TESLAっていう会社の人たちが、私たちの活動をサポートしていろいろ動いてくれてるの。地球では、こういう活動ってお金持ちからの寄付で賄うことが多いみたいだけど、私たちにはその概念がなくって。私たちの星では、お金持ちっていうのは大体、看護師さんとかバスの運転手さんとか、いわゆるエッセンシャルワーカーの人たちだしね。

CM: 地球が綺麗になったら、誰に1番そのことを伝えたいですか?

C: 地球のみなさんが、綺麗なところを自分たちで感じてくれれば。そして、自分たちが住んでる地球っていうのは本当に外から見ると青くて美しい星なんだって思ってほしい。あ、でもレイチェル カーソンさんにはちゃんと伝えたいな。私たち、お手紙も送っているしね。それから、ジョン レノンさん。あとは元アメリカ副大統領のアル ゴアさん。いろいろ勉強させてもらったし、私たちにとってもすごく大きな存在。

CM: 今後の予定は?

C: 活動資金は、自分たちでちゃんと稼いでいきたい。だから私たちのぬいぐるみも作って売っていくし。今後は、ファッションの人たちといろんなコラボもしてみたい。エコ、地球温暖化の阻止につながるようなもの、が大前提だけど。そうやって稼いだお金で、堂々と活動していきたいな。

**CM: What do you think about the situation around you now?**

C: I feel that the SDGs are very people-centred. Everything is labeled as harassment in the world today, don't you think? Even in your country of Japan, anything becomes harassment, from domesticated pets, zoos, aquariums, botanical gardens, and so on. People have all these big ideas about education, but in the end, it's all about the human ego. If you want to see penguins, spend your money and go to Antarctica or the North Pole to see them, rather than going to the zoo.

**CM: Doesn't that make you dislike people on Earth?**

C: Of course not. Your desires are what drives creations. People are the only creatures on Earth that can put themselves in other people's shoes. I still have hope, and I think that if everyone were to be kind to each other, the world would be a little more enjoyable. Conflict only causes sadness and hatred.

**CM: A planet without sadness and hatred would be great.**

C: Our sadness turned to anger which made us transform into what we are now. We're teaming up with our good friends at this magazine called commons&sense (this magazine) run by a company called CUBE, and this company NIKOLA TESLA is going to support our activities by organising events and PR. On this planet, these kinds of activities are usually funded by donations from rich people, but we don't want to rely on that concept, because on our planet, nurses, bus drivers, and other essential workers are the wealthy ones.

**CM: When the Earth becomes clean again, who would you want to tell it to first?**

C: The best would be if people on Earth can feel the change themselves. We'd like them to realise the beauty of their blue planet. Oh, but also, we want to tell Rachel Carson. We've been sending her letters, as well as to John Lennon. Al Gore, the former Vice President of the United States, has taught me a lot about the planet and has made an impact on my perspective.

**CM: What are your plans from here on?**

C: We want to earn our own money to fund our activities. That's why we're going to make and sell stuffed animals. Eventually, we want to collaborate with people in the fashion industry with an emphasis on being eco-friendly and preventing global warming. We can be proud to use the money we've earned to work on more projects to better the planet.

# YOU'LL NEVER FIND A RAINBOW IF YOU'RE LOOKING DOWN

*photos_Takuya Uchiyama fashion_RenRen hair_Shuco @3RD*
*make up_Akiko Sakamoto using for M·A·C COSMETICS @SIGNO*
*models_Shosei Ohira, Ren Kawashiri, Takumi Kawanishi, Syoya Kimata, Sukai Kinjo, Junki Kono, Keigo Sato, Ruki Shiroiwa, Shion Tsurubo, Issei Mamehara & Sho Yonashiro from JO1 @LAPONE ENTERTAINMENT*
*hair assistant_Takako Koizumi location_Glace Liberty Onpamido*

*all items by LOUIS VUITTON*

DO YOU WANT TO SEE MORE ?

teddy bear

jacket, shirt, sunglasses & necklace in hand

coat, shirt, sunglasses, bag, key rings attached to the bag & sneakers

hooded jacket, shirt, pants, sunglasses, necklace & bag

LV

top, shorts, cap, sunglasses, necklace, boots, bag & key rings attached to the bag

coat, shirt, pants, sunglasses & bag

cat

coat, shirt & sunglasses

coat, shirt, sunglasses, backpack, bag & sneakers

shirt, sunglasses & bag

shirt, pants, cap & bag

jacket, shirt, pants, necklace worn as a headpiece, sunglasses, bag & shoes

coat, sunglasses & backpack

奇跡を起こせ!!

Go to the Top!
“One for all, all for one”

Believe yourself.

空ちゃん
負けんなよ!!

Go for it!!

Be anything,
be yourself.

101 いず
負けんなよ!!

木全。

これからもよろしくお願いします!

影ら応援しとおり。

走れ、迷わず。

あなたは怒られた時どうしますか?
イライラした時はさぁやろう。
「新・基礎ウェーブ・改」

# CIAO! FALINE TOKYO

photos_Tawara

DO YOU WANT TO SEE MORE ?

2004 ヴァレンタインデイ オープン
2021 ヴァレンタインデイ クローズ

愛に溢れまくった原宿竹下通り。
17年分それぞれのFALINE LOVERSのドラマがあり、
ストーリーがあり、ヒストリーがありの第1幕フィナーレ。
私のお店だと思っていたら、
みんなのお店でした。

オールドスクールは懐かしく、
ニュースクールは新しく、仲間たちがフュージョンして
行く様を見て感無量。完全燃焼。ありがとう。

笑顔、涙、シャンパンで未来に乾杯。
FALINE TOKYOは
原宿の伝説になりました。

第2幕は MISS FALINE 北青山にて
繰り広げられます。

この想い出を残しておく機会をくれた
編集長佐々木さん & コモンズアンドセンスに感謝します。

Babymary xx

2004 Valentine's Day Open
2021 Valentine's Day Close

It was full of LOVE on Takeshita Street.
17 years FALINE Lovers Story & History
The finale was fantastic & Dramatic!
I thought the shop was mine,
but it was Everyone's.

Old School kids felt nostalgic...
New School kids were so fresh!!!
It was so beautiful, the whole family mixed together.
Thank you everyone who loved FALINE TOKYO.

Smiles! Tears!
Cheers for the future!
To be continued at MISS FALINE!

Special Thanx to Mr. Sasaki & commons&sense.

Babymary xx

Ciao!
FALINE
TOKYO

Ciao!
FALINE
TOKYO

I ♥
FALINE

Ciao!
FALINE
TOKYO

I
FALINE

Ciao!
FALINE
TOKYO
FALINE
4BLK
Thank you
FALINE
TOKYO
LISTEN
FLAVOR
ZENMALL TAKESHITA SHOP NO.5
SEARCH
AND
DESTROY
DROPS

Ciao!
FALINE
TOKYO
FALINE
tokyo.
DECOチョコ
FALINE
TOKYO

LISTEN
FLAVOR
FALINE
TOKYO
LOVE FALINE

FALINE
TOKYO

Asahi

FLAVOR

Faline Tokyo
Foever

LISTEN
FLAVOR

FALIN
TOK

Faline
KIMS
CIAO
FALINE TOKYO
HAPPY
VALENTINE'S DAY
TOKYO

# IMAGINATION MEANS NOTHING WITHOUT DOING

*photos_Takuya Uchiyama fashion_RenRen hair_Kunio Kohzaki @W*
*make up_Akiko Sakamoto using for M·A·C COSMETICS @SIGNO*
*models_Rintaro & Daiki Kanechika from EXIT @YOSHIMOTO*
*hair assistant_Aiko Pink Tanaka background photos_AFLO*

*all items by DIOR*

DO YOU WANT TO SEE MORE ?

coat, shirt, pants & scarf
earring MODEL'S OWN
location_Grand Socco, Morocco.

jacket & tank top
location_Koutoubia Mosque, Morocco.

FROM LEFT:
coat, shirt, pants & scarf
earring MODEL'S OWN

jacket, tank top & pants

location Windhoek, Namibia

shirt, top, shorts & belt
earring MODEL'S OWN
location Fez, Morocco.

shirt, top, pants, belt & bag :
**location_Marrakesh, Morocco.**

FROM LEFT
shirt, top, shorts, belt & socks,
earring MODEL'S OWN

shirt, top, shorts, belt & socks

**location_Grand Socco, Morocco.**

sweater, top, pants, belt, socks & sandals
location_Cape Town, South Africa.

sweater, top, pants & belt
earring MODEL'S OWN
location_Cape Town, South Africa.

sweater, top, shorts, beret, bag, socks & sandals
**location_Rabat, Morocco.**

jumpsuit, sweater, beret & belt

I PAY NO ATTENTION WHATEVER TO ANYBODY'S PRAISE OR BLAME. I SIMPLY FOLLOW MY OWN FEELINGS.

*photos_Yuji Watanabe fashion_Shino Itoi*
*make up_Akiko Sakamoto using for M·A·C COSMETICS @SIGNO*
*model_Roy @TWIN PLANET photo assistant_Ryohei Hashimoto*
*background photos_AFLO*

*all items by GUCCI*

DO YOU WANT TO SEE MORE ?

hoodie, pants, glasses & socks
location_Luna Park in Coney Island, USA.

shirt, shorts, knit cap & belt
location_Germany

polo shirt, shorts & sunglasses
location_Luna Park in Sydney, Australia.

shirt, shorts, pants, sunglasses, earring,
necklaces, bracelets & rings

location_Nottingham Goose Fair, U.K.

jacket, pants, earrings, bag & shoes
location_Luna Park in Coney Island, USA.

t-shirt, pants, bag, hat, earring, belt & sneakers_

location_London, UK.

cardigan, polo shirt, pants, knit cap, sunglasses, earrings, necklace, rings, bag & sandals
location_Luna Park Funfair in Scarborough, UK

cardigan, polo shirt, knit cap,
sunglasses, earrings & necklace

t-shirt, bag, earring & bracelet
location_Boardwalk
in Coney Island, USA.

# A TRAMP, A GENTLEMAN, A POET, A DREAMER, A LONELY FELLOW, ALWAYS HOPEFUL OF ROMANCE AND ADVENTURE.

*photos_Yume Ippei fashion_RenRen hair_Kunio Kohzaki @W*
*make up_Akiko Sakamoto using for M·A·C COSMETICS @SIGNO*
*model_Rihito Itagaki @STARDUST PROMOTION*
*hair assistant_Sachi Oohara/background photos_AFLO*

*all items by VALENTINO*

location_One World Trade Center, USA.

jacket, shirt, shorts & sandals
**location_Dubai Marina, UAE.**

jacket, shirt, jeans & sneakers

**location_Shanghai World Financial Center, China.**

hooded top

blouson, t-shirt, pants, bag & belt
location_New York City, USA.

shirt & bag

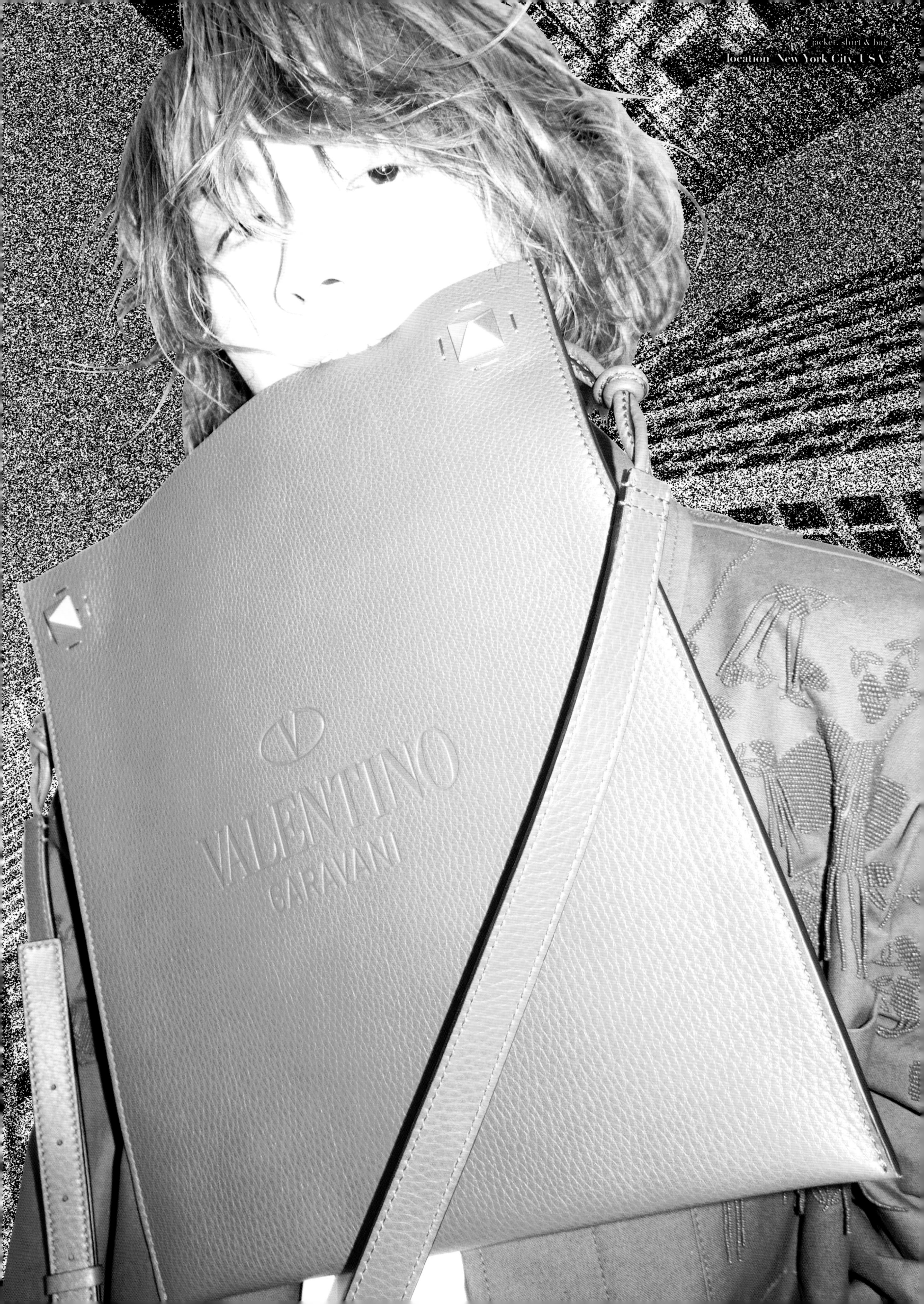


jacket, shirt & bag
location_New York City, U.S.A.

shirt & bag
location_Singapore.

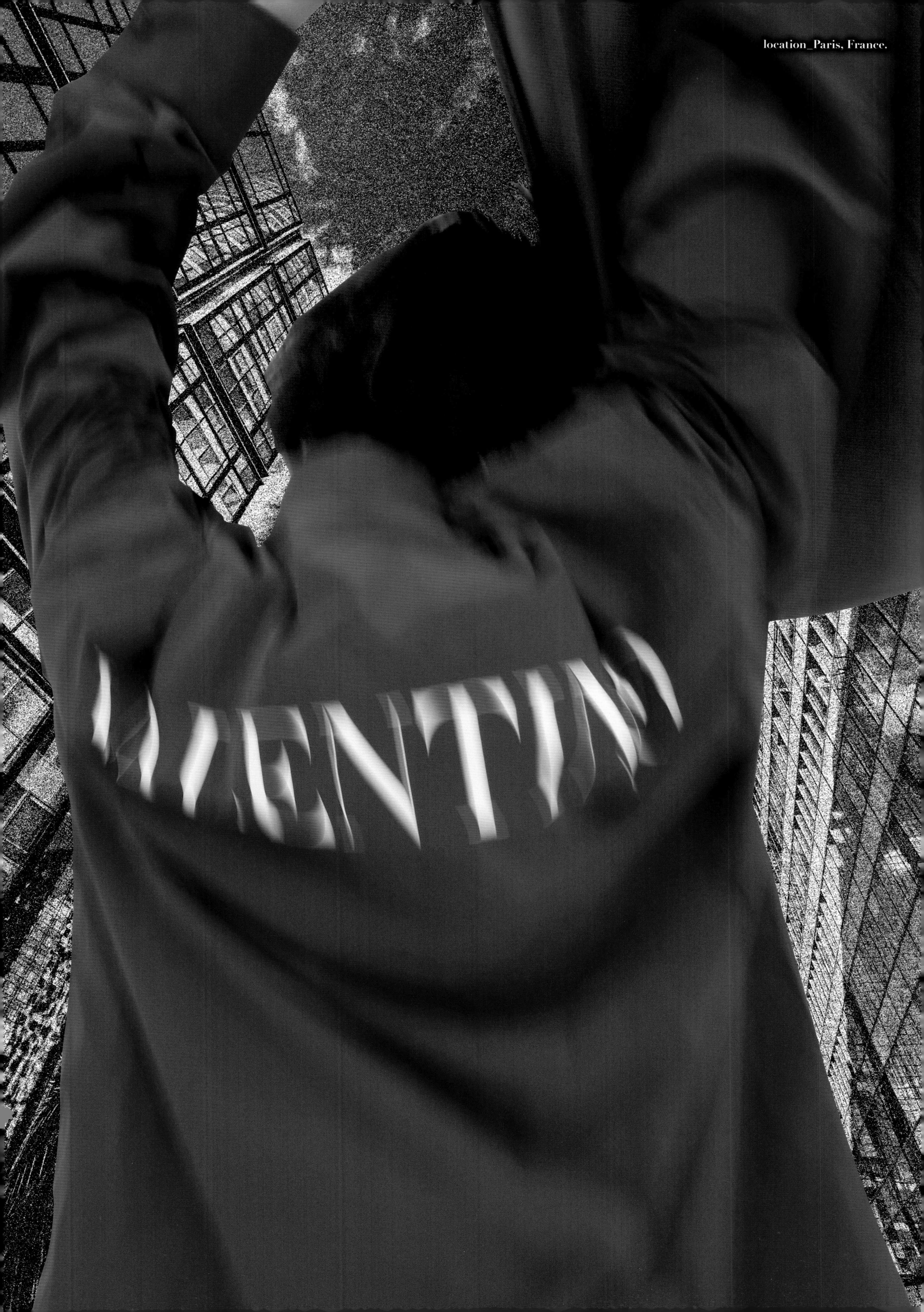

location_Paris, France.

DO YOU WANT TO SEE MORE ?

TOKYO DANDY - SCENE 16
SIGN OF THE TIMES
photos & text_Tokyo Dandy
Empty billboards.
A sign of the times?
Or a clean slate? A place and space to project our own thoughts and ideas.
Spring - a time to start afresh.
Now more than ever.
Who do you want to be?
PROFILE: TOKYO DANDY was established in 2008 to bridge the gap between street and high fashion. In the 3 years since then founding members Dan and Joe have gone on to dominate the Tokyo party and fashion scene defying genre boundaries in mainstream Japanese culture to bring together the worlds of fashion, art, music and culture and lead a new creative generation.

TOKYO DANDY - SCENE 16
SIGN OF THE TIMES
photos & text_Tokyo Dandy
空のビルボード。
これは時代の象徴?
それとも、白紙に戻ったのか?それは、自分の考えやアイディアを投影する場所と空間。
春は新たなスタートを切るとき。
今こそ、これまで以上に。
あなたは誰になりたい?
PROFILE: ハイファッションとストリートファッションの架け橋として、2008年にTOKYO DANDYを設立。その後3年間でファウンダーであるダンとジョーは東京のパーティ&ファッションシーンを支配するまでの存在に。日本のメインストリームカルチャーと、ファッション、アート、カルチャーというあらゆるジャンルの間に存在する壁を破り、新しいクリエイティブ ジェネレーションをリードする存在。

# EVERYTHING SHOULD BE MADE AS SIMPLE AS POSSIBLE, BUT NOT SIMPLER.

photos_Naohiro Tsukada props_RenRen photo assistant_Senta Murayama

all items by **VITAL MATERIAL**

DO YOU WANT TO SEE MORE ?

FROM TOP:
**AROMA HERB BATH SALT** /
aqua shine
clear sky
healing rose

VITAL MATERIAL
organic
Sel de bain bath salt
e350g→12.3floz

FROM LEFT:
AROMA DIFFUSER E.D.T VER /
un
deux

AROMA DIFFUSER /
damask rose
bergamot

VITAL MATERIAL
diffuseur de parfum d'ambiance
MATERIAL

FROM TOP LEFT:
MULTI DELICATE DETERGENT /
deep green ocean
floral & pear bouquet
royal jasmine tea

FABRIC CONDITIONER /
deep green ocean
floral & pear bouquet
royal jasmine tea

SUBLI
DESCRIPTION
VITAL MATERIAL
SUBLI
FABRIC CONDITIONER
FLORAL & PEAR BOUQUET
450ml / 15.2 fl.oz.
SUBLI
FABRIC CONDITIONER
ROYAL JASMINE TEA
450ml / 15.2 fl.oz.
DESCRIPTION
A SPECIAL LAUNDRY DETERGENT BRAND CREATED TO RETURN MATERIALS TO THEIR ORIGINAL CHARACTERISTICS AND ALLOW THEM TO BE WORN LONGER WITH THE AIM OF REALIZING ANTI-AGING CLOTHING. AS A MATERIAL THAT IS USED NOT ONLY FOR CLOTHING BUT ALSO FOR COSMETICS AND IS FRIENDLY TO HUMANS AND THE ENVIRONMENT, WE CAREFULLY SELECT ONLY PLANT-DERIVED AND HIGHLY SAFE INGREDIENTS. USE : WOOL / SILK / COTTON / SYNTHETIC FIBER COMPONENT : WATER / QUATERNARY AMMONIUM COMPOUNDS / POLYOXYE THYLENE ALKYL ETHER / ETHYLENE GLYCOL / INORGANIC SALT / FRAGRANCE / MATRICARIA / SAGE / JOJOBA / ROSEMARY / MADE IN JAPAN
VITAL MATERIAL

ROOM FABRIC MIST / floral & pear bouquet

AND
AND VITAL MATERIAL FIL DENTAIRE #201 CITRON
WORLD IS A BOOK AND THOSE WHO DO NOT TRAVEL READ
ONLY ONE PAGE. AND ORGANIC. AND FASHION.
AND DESIGN. AND PLAY. AND FOR YOU.
&
VITAL MATERIAL

DENTAL FLOSS / lemon

AND
AND VITAL MATERIAL
THE WORLD IS A BOOK AND THOSE WHO DO NOT TRAVEL READ ONLY
ONE PAGE. AND ORGANIC. AND FASHION. AND DESIGN.
AND PLAY. AND FOR YOU.
#101 HERBES MENTHE
BAIN de BOUCHE 360ml / 12.1 fl.oz.
&
VITAL MATERIAL

MOUTHWASH / herb mint

AND
VITAL MATERIAL DENTIFRICE BLANCHISSANT #101 HERBES M
THE WORLD IS A BOOK AND THOSE WHO DO NOT TRAVEL READ ON
AND ORGANIC. AND FASHION. AND DESIGN. AND PLAY. AND F
AND
ENTIFRICE SOINS BUCCAUX #102 MENTHE CITRON 75g / 2.6oz.
OOK AND THOSE WHO DO NOT TRAVEL READ ONLY ONE PAGE.
AND FASHION. AND DESIGN. AND PLAY. AND FOR YOU.

FROM LEFT:
TOOTHPASTE /
lemon mint
herb mint

MEGA
BLD.
IMAGINATION IS MORE IMPORTANT THAN KNOWLEDGE. KNOWLEDGE IS LIMITED. IMAGINATION ENCIRCLES THE WORLD.
photos_Takuya Uchiyama
fashion_Shino Itoi
grooming_Go Utsugi @PARKS
model_Shuhei Uesugi @KEN ON
background photos_AFLO
all items by SAINT LAURENT
DO YOU WANT TO SEE MORE ?

jacket, shirt, jeans, sunglasses, bracelet, belt & shoes

*location_7 cho-me Ginza, Japan.*

shirt, pants, sunglasses, choker, necklace, belt & bracelets
location Shibuya Center-Gai, Japan

shirt, t-shirt & necklaces

shirt (outside) & shirt (inside) , pants, necklace
location_Dotonbori, Japan.

cardigan, shirts, sweater, pants, sunglasses, necklace & bracelet

食べ放題
2,480円

blouson, shirt, pants, choker, necklace & sandals

# THE SHORTER WAY TO DO MANY THINGS IS TO DO ONLY ONE THING AT A TIME

*photos_Yume Ippei fashion_Shino Itoi*
*grooming_Kunio Kohzaki @W*
*model_Kohei @BE NATURAL*
*grooming assistant_Sachi Oohara*

*all items by PRADA*

Starter
SINCE 1971

jacket & shirt

PRADA

coat

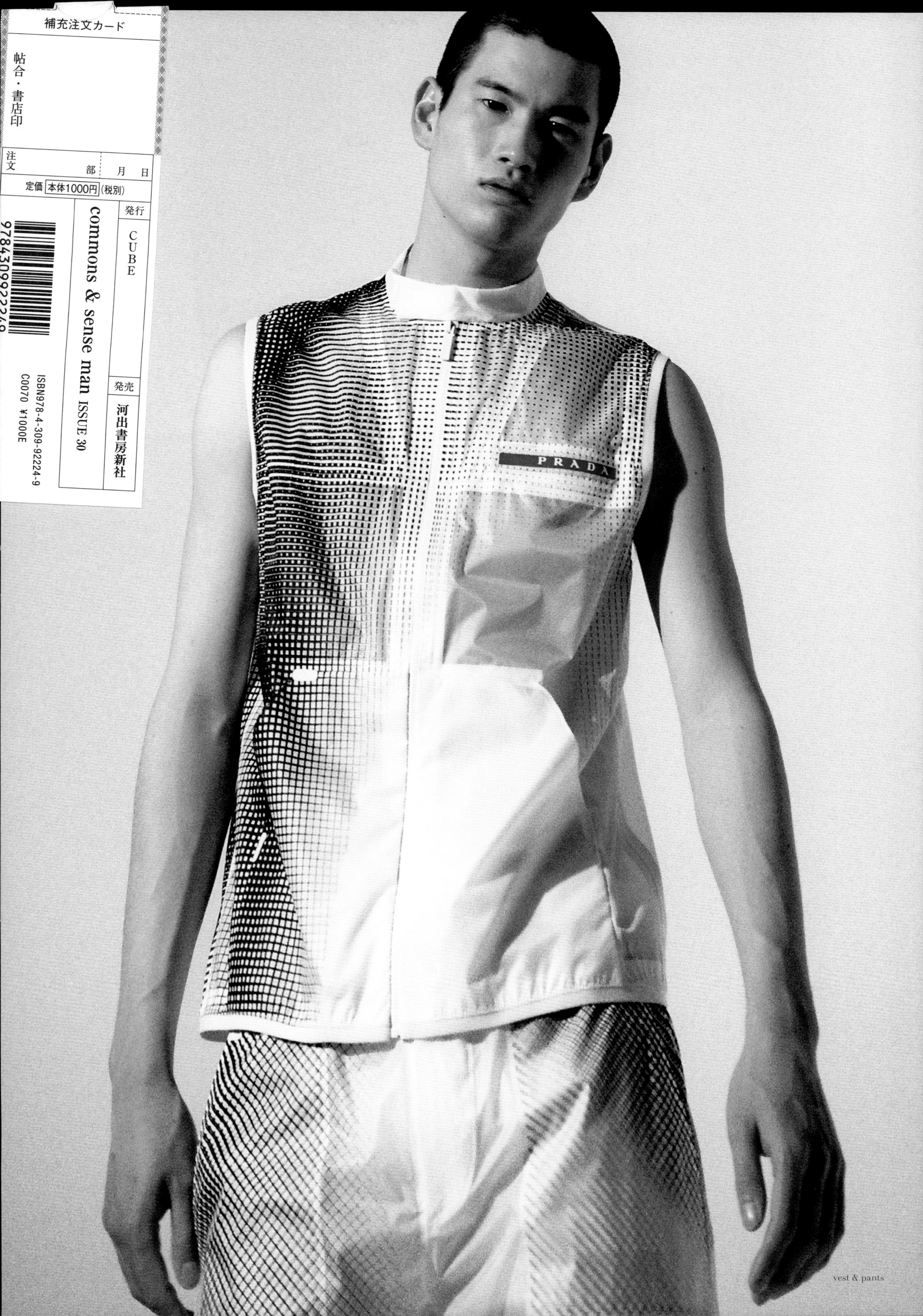
補充注文カード
帖合・書店印
注文 部 月 日
定価 本体1000円(税別)
発行 CUBE
commons & sense man ISSUE 30
発売 河出書房新社
ISBN978-4-309-92224-9
C0070 ¥1000E
PRADA

vest & pants

販売調査カード

commons & sense man ISSUE 30

このカードは、河出書房新社営業部売上カード係宛に定期的にお送り下さい。貴店の販売実績として、今後の配本等に活用いたします。

〒151-0051
東京都渋谷区
千駄ヶ谷2-32-2
㈱河出書房新社

定価 本体1000円（税別）

BBBN4-309-92224-4 C0070 ¥1000E

shirt

bag & tie

jacket, shirt, pants, socks & shoes

ANYONE WHO
HAS NEVER MADE
A MISTAKE HAS
NEVER TRIED
ANYTHING NEW


*photos_Yume Ippei fashion_RenRen hair_Kunio Kohzaki @W*
*make up_Akiko Sakamoto using for M·A·C COSMETICS @SIGNO*
*models_Nico, Kai de Torres & Kio de Torres from GLIHCO,*
*Kaiga Nakamura @STANFORD, Kazane, Yuto Kubota, Kagiru @DONNA*
*photo assistant_Pao hair assistant_Sachi Oohara*

*all items by DIOR*

*DO YOU WANT TO SEE MORE ?*

coat, top, pants, gloves, belt, bag & boots

coat, shirt, top, pants, beret, bag & boots

coat, shirt, top, pants, bag & sneakers

coat, top, pants, belt & boots

sweater, top, pants, earring, bag, belt & boots

jacket, top, pants, bag & boots

coat, blouson, top, pants, belt, bag, socks & slippers

OUR RICHES,
BEING IN
OUR BRAINS.
photos_Kenichi Yoshida fashion&props_RenRen
illustrations_Masanori Ushiki
all items by LOUIS VUITTON
DO YOU WANT TO SEE MORE ?

LOUIS VUITTON
死
正義
愛

LOUIS VUITTON PARIS
LOUIS VUITTON PARIS
LOUIS VUITTON PARIS

2021

# INTELLECTUALS SOLVE PROBLEMS, GENIUSES PREVENT THEM.

photos_Kenichi Yoshida  fashion&props_RenRen

all items by **SAINT LAURENT**

TOTE BAG **SAC DE JOUR SUPPLE** / black

SAINT LAURENT

BAG **SAC DE JOUR SUPPLE** / brown

RIVE GAUCHE
SAINT LAURENT

RIVE GAUCHE
SAINT LAURENT

BAG **NUXX** / brown & black

BAG **CAMP** / beige, kahki & black

SNEAKERS **MALIBU** / black & off white

TOKYO CLOCK

SAINT LAURENT
PARIS
SAINT LAURENT
PARIS

SNEAKERS **SL24** / multicolour

SAINT LAURENT

# LIFE IS A TRAGEDY WHEN SEEN IN CLOSE-UP, BUT A COMEDY IN LONG-SHOT.

photos_Kenichi Yoshida fashion&props_RenRen

all items by **VALENTINO GARAVANI**

DO YOU WANT TO SEE MORE ?

VALENTINO
GARAVANI

brown

BAG **IDENTITY** / brown

**ATELIER BAG** / white & red

**ATELIER BAG** / white & red

**ATELIER BAG** / white & red

**ATELIER SHOES** / white

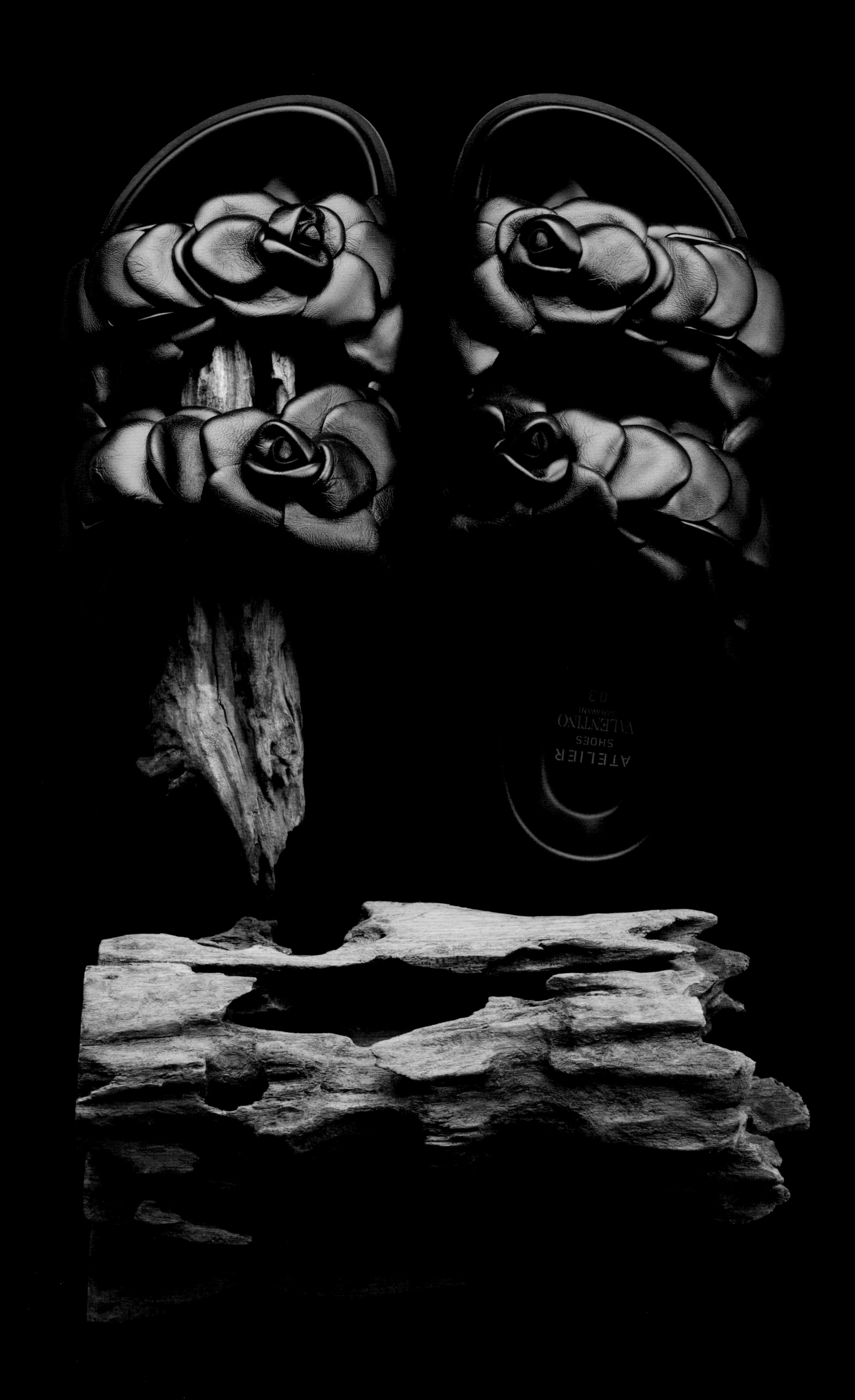

**ATELIER SHOES** / brown

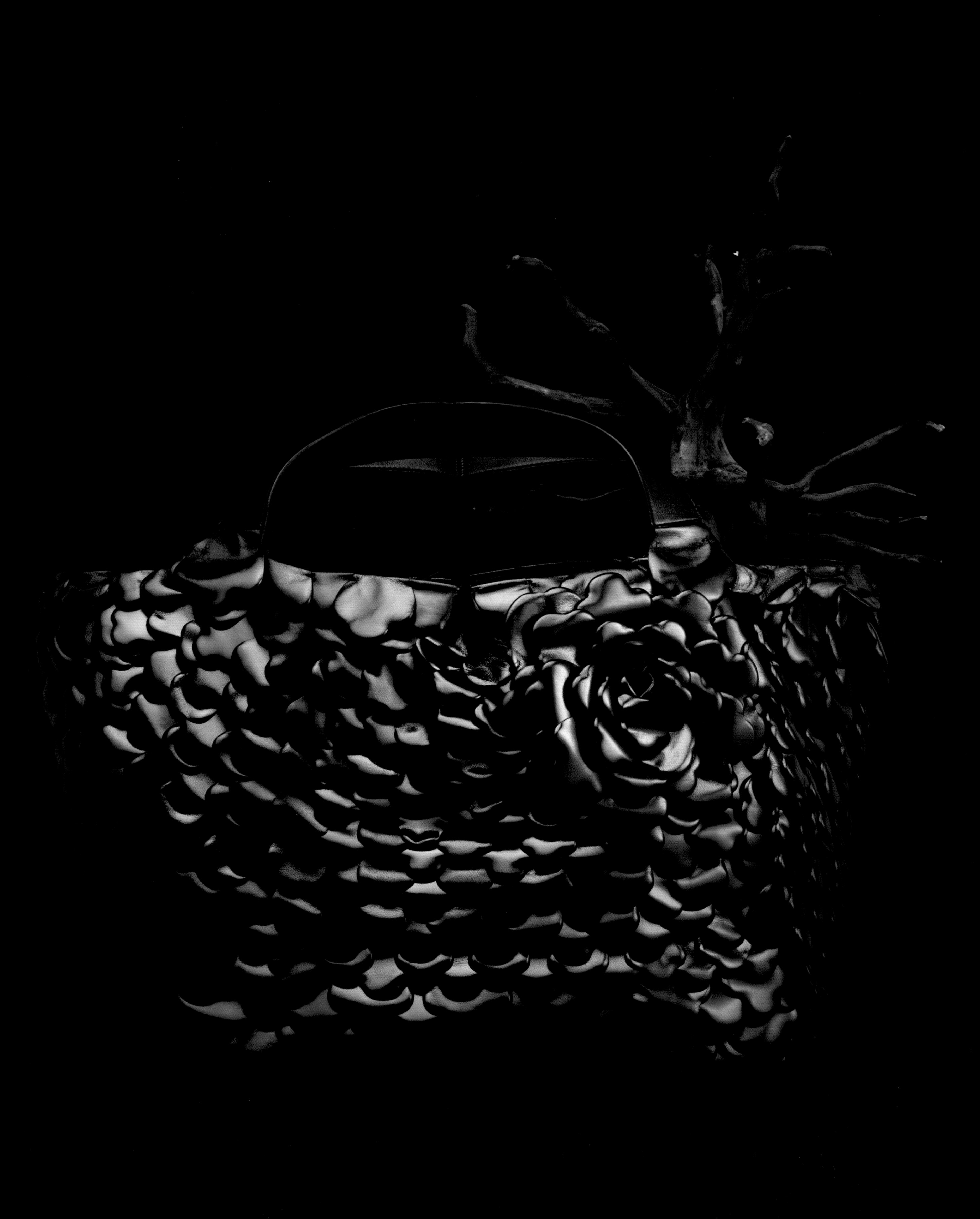

**ATELIER BAG** / black

**MACRAME SNEAKERS** / black

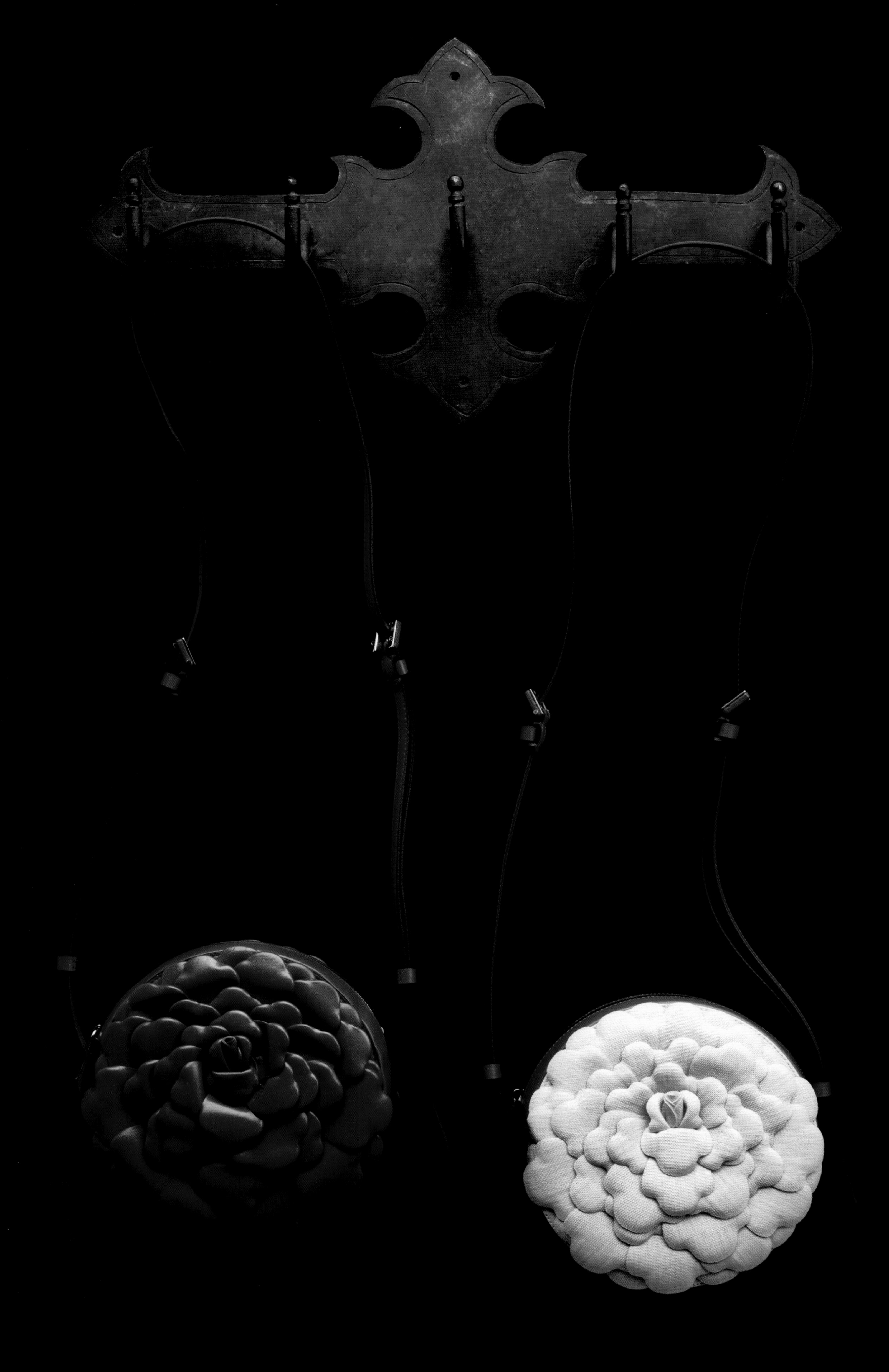

FROM LEFT:
**ATELIER BAG VG** / red

white

# LEARN FROM YESTERDAY, LIVE FOR TODAY, HOPE FOR TOMORROW. THE IMPORTANT THING IS NOT TO STOP QUESTIONING.

photos_Kenichi Yoshida fashion&props_RenRen

all items by **DIOR**

DO YOU WANT TO SEE MORE ?

**CUMMER BELT** / white, red & yellow
flag_Republic of Ghana

**MINI "SADDLE" MESSENGER** / beige & black
flag_Republic of Rwanda

**MINI "SADDLE" MESSENGER** / black
flag_Republic of Tunisia

**MINI "SADDLE" MESSENGER** / black

flag_Kingdom of Morocco

**MAXI “SADDLE” BAG** / beige & black
flag_Republic of Malawi

**MAXI "SADDLE" BAG** / black
flag_Republic of Angola

**RIDER BACKPACK** / navy
flag_Republic of Namibia

**CROSSBODY POUCH** / navy
flag_Republic of Equatorial Guinea

**DIOR LOCK MINI BAG** / black
flag_Republic of Senegal

**DIOR LOCK MINI BAG** / beige & black
flag_Republic of Guinea

**DIOR ATLAS SANDALS** / beige, black & navy

**DIOR ATLAS SANDALS** / navy, black & beige

**"B23" HIGH TOP SNEAKERS** / white & yellow

**“B27” MID-TOP SNEAKERS** / grey, white, navy, blue & red

# MARCH 11th

**photo_LOOM NIPPON　english translation_Jordan A. Y. Smith**

## 『嗚呼　三月十一日』

加賀美 誠・友吉 鶴心 / 共作

我 日の本は 自然(ジネン)災い 多き國
わけても 大いなる震る事も
幾あまた
又その悲しみも
星の数程 幾あまた
忘れまじ 嗚呼 あの日 あの時を

まだ春寒き 3月11日 斑雪
時刻は午後二時46分18秒

宮城県牡鹿半島東南東の沖合い
130キロより
ごお ごお と鳴り渡る 地の叫び
おびただし
足 すくわれんばかりの 大なゐ震る
恐怖なる 大なゐ震る

はた と止みたる 静けさは
空気が凛と張りつめん時も
とまらんばかり
不気味なほど に 静かなり

ほどなく
響きわたるはのろしのごとき
緊急無線

海立つ(ツナミ)
海立つ(ツナミ)が来くるぞ

聞けば
一早く高所に逃ぐる人あれば
恐怖におののき 踞り言葉なき人
また 散乱したるを整えんと
様々に

重ねて 流れる
緊急無線の遠き先には
水平線から迫りくる海面(カイメン)は
遥かに高き漆黒の壁

音なき黒き壁は岸を目指してし
迫りくる

その恐ろしき有様を目前に
逃げもせず　ただ人々の為に
声を叫びの 緊急無線

逃げろ 逃げろ 逃げろ
黒き壁はその声も町と共に呑みつくす

黒き冷たき大浪は
浸々と瞬くあいだに
湊をはじめ 人々の家 人々の店 人々の車　総てを押し流し
更なる壁となり
更に総てを押し流す

人々の轟に怯ゆる声も
人々の逃げろ 逃げろの絶叫も
うち消され
黒壁は総てを呑み引くその有り様は
夢も希望も幸せも
何もかも大海原に連れ去りて
あと志津川歌津の有り様は
ただ 黒壁の灰塵と化す

早 大魔が時も過ぎ行きぬ
町はなくただ夢かと
静々寂々 ただ夢かと
いつしか雪も晴れわたり
冷たき空にはただ 星 一つ
また星一つ　星一つ
現の星は悲しみの極みかと

## "March 11th"

co-production_Makoto Kagami & Kakushin Tomoyoshi

Land of our motherland and the land of rising sun.
land where natural disasters abound,
Above all, the enormous quakes,
So many of them,
The sadness as infinite as stars in the sky,
Unforgettable, alas, that day, that moment

Spring chill, March 11, lingering patches of snow,
The time was 14:46:18

Miyagi Prefecture, Oshika Peninsula, East-southeast
from 130 km offshore,
The roaring arrived,
the roar of the land,
An enormous trembling,
inescapable, an infinite,
terrifying stampede

Suddenly pausing, the silence,
tension pregnant air fit to stop
The silence borders on eerie....

Soon enough it comes resounding like a starting gun,
Radios blare the emergency

The ocean rises: tsunami
The tsunami is coming

Anyone listening from high up would hear it even quicker,
Shudders of fear, cowering people bereft of words,
Again, scattered confusion,
need to get it together before all falls apart

Again, the broadcast,
Emergency signals from the other side of the radio,
From the horizon, the ocean surface now a pitch black wall,
looming, closing in

The soundless black wall aiming straight for the cliffs

A horrific spectacle unfolds before my eyes,
Not fleeing, for everyone's sake,
The radio blares its emergency warnings

Evacuate! Escape! Run!
The black wall engulfs that voice along with the town

The giant, black, frigid wave
Soaking everything in a blink,
First the harbor, then people's houses, people's shops, people's cars, all is washed away
Forming yet another wall,
washing away everything in its path

People's voices, yelling or quavering,
People's screams, Run! Run!
Smashed into oblivion,
With the black wall swallowing everything,
Dreams and hopes and happiness, all is washed away into the ocean depths,
Then Shizugawa and Utatsu--
reduced to rubble by the black wall

Already, a great demon has frozen the flow of time,
No town left, is this a dream?
Silence, stillness, is this a dream?
The snow stops falling,
Frigid air, and one lone star,
Then one star after another appears,
the saddest thing ever.

※現在、作品作曲中なので今後歌詞の変更もあります。
*The lyrics may change in the future as the work is currently being composed.

DO YOU WANT TO SEE MORE ?

# THE VALUE OF A MAN SHOULD BE SEEN IN WHAT HE GIVES AND NOT IN WHAT HE IS ABLE TO RECEIVE

**interview with SASKIA LENAERTS, ALECSANDER ROTHSCHILD, ALEX WOLFE & DINGYUN ZHANG**

**photos_Matteo Carcelli**

BIRKENSTOCKがセントラル セント マーチンズ ファッションヒストリー＆セオリーコース及び、ファッションコースと共同し始動した初の教育プロジェクト、BIRKENSTOCK×CSM。このプロジェクトでは、セント マーチンズの学生らがBIRKENSTOCKのアーカイブのリサーチを元に、アイコニックなスタイルをモダンに再解釈し、それぞれ自分たちの手がけるデザインに落とし込んで発表した。デザインは、BIRKENSTOCK CEOのオリヴァー ライヒェルト氏、ファッショコースMAのディレクター ファビオ パイラス、そしてサラ モワーなどの12名で構成される審査員によって審査され、ファイナリスト10名が選出された。そのなかから最終的に、サスキア レナエルツ、アレクサンダー ロスチャイルド、アレックス ウォルフ、ディンユー チャンの4名が手掛けたデザインが実際に生産化され、2021年より発売開始となる。今後のファッション界を担うエネルギーに溢れた次世代のデザイナーである彼らに今回のコラボレーションの話を聞いてみた。

**BIRKENSTOCK × CSM is the first educational project launched by BIRKENSTOCK in collaboration with Central Saint Martins Fashion History & Theory and Fashion courses. In this project, St. Martin's students researched BIRKENSTOCK's archives to create a modern reinterpretation of the iconic style, which they then incorporated into their own designs. The designs were judged by a 12 jury members, including BIRKENSTOCK CEO Oliver Reichert, Fashion Course MA Director Fabio Piras, and Sarah Mower. 10 finalists were selected. The final four designs, by Saskia Lenaerts, Alecsander Rothschild, Alex Wolfe, and Dingyun Zhang, have been put into production and will be available for purchase in 2021. We asked these 4 students, who are the next generation of energetic designers who will lead the fashion world in the future, about this collaboration.**

ommons&sense man (以下CM)：今回のプロジェクトに参加して、どのような経験
されましたか？
スキア レナエルツ(以下SL)：デザインプロジェクトに留まらず、最後まで商品開発に
わる機会を得ることができました。ケルンに行って自分たちでプロトタイプを作った
、製品開発者と一緒に仕事をして彼らの知識から学ぶことができたのでやりがいが
りました。

レクサンダー ロスチャイルド(以下AR)：デザインアイディアを出してから実際にケル
にある生産工場に行き、検証しながらより具体的なものにしていくというプロセスを
て製品を作ることは、非常に大きな経験でした。

レックス ウォルフ(以下AW)：本当に刺激的でした。大学の授業とは別のこのプロジェク
に参加できたことは、私たち若手デザイナーにとってとても重要でした。このような業
での経験を積むことができるのは本当に素晴らしいことです。

ィンユー チャン(以下DZ)：BIRKENSTOCKとのコラボレーションでは、制作チームと密
連携してデザインを実現させました。制作側とのコミュニケーションが実験的なアイ
ィアを商品化することに本当に役立つことを学びました。

M：デザインのプロセスについて教えてください。
：BIRKENSTOCKのフットベッドはブランドの基本であることを知ったので、その伝統
敬意を表しながら、よりオーガニックでやわらかい流動的なものを作りたいと思いま
た。上部分のレザーの間にレイヤーを挟み、トップステッチでキルティングを施し、
レガントで女性らしいものに仕上げています。重視したのは、ストラップの上からでは
く、ソールの横顔が全体的に見えるようにしたことです。遠くから見たときに、有機的
滑らかなラインが一周しているように見えるようにデザインしました。

R：私の場合は、家族の歴史を振り返ることからはじめました。代々、芸術家、画家、彫刻
などの家系で、特に祖母の好きな彫刻家ブランクーシからインスピレーションを得て
ます。靴の上部分はブランクーシの彫刻のようなデザインになっており、彼の視点や
品作りにも影響を受けています。シルバーのタイプはブランクーシの素材の使い方と

**commons&sense man** (hereafter CM): **What kind of experience was it fo you to participate in this project.**
Saskia Lenaerts (hereafter SL): Rather than just staying with the design project, yo got the opportunity to really follow through. Going to Cologne and working on th prototype ourselves, and working together with the product developers an learning from their knowledge… it was extremely rewarding to see how it really go to being something that can be commercially sold and then be appreciated.

Alecsander Rothschild (hereafter AR): Following the process of having the initia design ideas, actually going to Cologne to their production site, testing ideas an making it more tangible… This entire experience of making the product that going to be sold was really a huge experience.

Alex Wolfe (hereafter AW): It was really exciting. Being involved in this project whic took place outside the university was really important for us young designers because there's a real element to it. To have this kind of industry experience i really amazing.

Dingyun Zhang (hereafter DZ): During the collaboration with BIRKENSTOCK, w worked closely with the production team to bring our design to reality. I learnt tha communication with the production can really help generating experimental idea into commercial products.

**CM: Tell us about design process.**
SL: We learnt that the BIRKENSTOCK's footbed was so fundamental to what the do and it was the one element in every style which is the same. So I wanted t celebrate that heritage and create something more organic, less rigid or mor fluid. The upper leather has a layer sandwiched in between and it's quilted with to stitching to create something elegant and more feminine. What I personall emphasised was to see the entire profile of the sole run through, not have the strap go over. So that when you look at it from the distance, it's one smooth organic line going around.

AR: My idea initially was to look back on my family background. I come from a family of artists, painters and sculptors specifically. So I basically used m grandmother's favourite sculptor Brancusi as my main inspiration. The upper par is a shape taken from one of the Brancusi's sculptures while also being inspired b his way of looking and his way of making. The silver version is a contrast of how

タイプは、他のカラーの影を映しているバージョンだと思っています。上部分を取り外すこともできるので、よりシンプルにすることも可能です。

AW: サンダルは医療用のレッグブレースとモトクロスのシンガードを組み合わせたアイディアから着想を得ています。プロジェクトのサブタイトルは "break-a-leg"（直訳: 足を折るという意味だが、幸運を祈るという意味）。シリアスな部分と遊び心を掛け合わせ、皮肉でひねりを加えたデザインです。後ろ側の調節可能なマジックテープのストラップなど、医療用フットウエアから取り入れたディテールがたくさんあります。トップシールドを外してローシューズとして履くことも可能なデザインです。

DZ: 私のデザインのコンセプトは日常履きとしての親しみやすさと、現代のコマーシャルフットウエアデザインとの関連性です。要約すると、快適なデザインで、あまり考えすぎずに履ける、シンプルで履き心地の良い、汎用性の高い靴を作りました。パッドスタイルとボンドエッジがとてもシンプルで、靴底はレザーで覆われています。

**CM: 今、1番興味を持っているものは？**
SL: 政治のことを学んだり、世界情勢や異文化についてとても興味があります。皆が同じ未来を見ることができる方法を見つけたいと思っています。私にとってファッションは、それを実現するためのツール。人の見た目や着こなしは大事な第1印象なので、ファッションだけで自己紹介もせずに人のことを判断することができるのはとてもパワフルなこと。ファッションを通して人生をもっと楽にすることができると思います。

AR: 今は、"グラマラスさとダサさの境界線"に興味があって、ジョン ウォーターズの映画や、ディバイン（ドラッグクイーン）のインタビューをたくさん見ています。とても好きです！

AW: 今はマスキュリンの世界に遊び心を取り入れることに興味があります。私にとってとても重要なことで、次のコレクションでさらに追求していきたいと思っています。より多くの男性にファッションに興味を持ってもらえるように、刺激的で包容力のある男性のアイデンティティを作りたいです。私のスタイルは、平凡で真面目なものをひっくり返すことによって浮き出る奇妙な視点からきていると思います。エルヴィン ワームのように身体に彫刻を施すアーティストに興味を持っていて、その非常に瞬間的な表現方法を

standing on brass or silver. Another one is a completely black version. Basically the black version comes from the idea of having just a shadow of the other shoe You can also remove the upper part so it's more simple and also the general idea of the shoe was to make something simple.

AW: For me, I really wanted to do something that I enjoyed. The MOTO sandals are inspired by the idea of the medical leg brace, combined with the motocross shin guard. The subtitle for my project is "break-a-leg" , an ironic twist with something serious and something playful combined together. So, there are alot of details that I have taken from the medical footwear, such as the adjustable velcro straps on the back. The top shields are removable and can be worn as low shoes.

DZ: My concept is a close versatility with the everyday wear and its relevance with current commercial footwear design. So basically, it is the comfort which comes from the most comfortable place for us. The idea of wearing them on our feet is like…the simple, comfortable and versatile design that can be worn without too much thinking. In general, they're super simple shoes with the pad style and bonded edge, and the soles are covered with leather.

**CM: What are you most interested in at the moment?**
SL: I am very much into studying politics and what happens in our world an learning about different cultures…I want to address what I see as an ideal outcome and one disarmed, find a way where we can all lift together globally and locally and just accept each other. For me fashion is being really good outlet to address tha The way we look and the way we dress, is the first impression people get from us Without saying any words, nor introducing ourselves. I think that is very powerfu It is a very good tool to address those issues and life would be much easier.

AR: Right now, I am interested in the juxtaposition of glamour and bad taste. So have been watching a lot of John Waters' films and interview of "Divine" the drag queen. I love that!

AW: At the moment, I am interested in bringing a kind of sense of playfulness around masculinity. I am really interested in pursuing my next collection. I am interested i creating exciting and inclusive identity for men, which could interest more men i fashion. I have this aesthetic of taking something very mandane and all serious an turning it upside down, which gives a kind of weird perspective of things. I am interested in artists such as Erwin Wurm who works with sculptures on the body. I

DZ: 生分解性のあるナイロンや、防水性のある生地を開発することに興味があります。その素材を使って、将来的には若いデザイナーと一緒にコラボレーションをしたいです。今は工場と連絡を取り合って研究しているところです。また、軽量で生分解性のあるナイロンを、生産や機能的な輸送のために革新していきたいと考えています。

**CM: 次世代のファッションデザイナーとしての目標はありますか?**
SL: デザイナーになりたいと思っていますが、アーティストのように複数のプロジェクトを持って、今回のようにコラボレーションするのも良いですね。今のパンデミック時代においては、いかにして大量生産の成長を食い止めるかということに重点が置かれはじめていると思います。今の時代だからこそ、若いデザイナーとしても自分たちのやり方でやっていくチャンスがあると思うので、これまでのシステムを踏襲する必要はない。自分自身を信じて、やりたいことをやり続ける自信を得ることは大切なことだと思う。これからいろいろなことを経験して好きなことと好きでないことを見つけるのが楽しみです。

AR: 私が目指しているのは、まず第1に愛のあるものづくりです。ほとんどの人がクリエイションをはじめるとき、思いやりからはじめています。愛を意識することによってお互いへの接し方や生産の仕方など、業界のあらゆる部分が改善していくと思います。

AW: 私たちのような若手デザイナーにとって、今、自分の視点を表現することはとても重要なことです。大規模なブランドは、消費サイクルの中で真のメッセージが見えなくなってしまうことがあります。若手デザイナーは、ユニークなビジョンを追求すればするほど、より純粋で誠実な部分をコントロールすることができるので、私は若いクリエイティブな人たちが自分たちの個性を追求していくことをサポートしたいです。その観点がクリエイティブ業界にもたらす影響はかけがえのないものだと信じています。

DZ: 私は今後も素材の研究を続けていきたいと思っています。私はコレクションを発表していますが、つねに生分解性の解決策を追求していますし、生産量が少なく消費者にとってコストのかからない衣服を作るための生地の戦術を追求しています。

DZ: I am interested in developing my own type of biodegradable nylon and waterproof fabrics at the moment. I wish to use these materials to collaborate wit the young designers in the future to work together. Right now, I'm just doing alot o research and contacting factories, promoting... I also plan to innovate light weigh biodegradable nylon for production from elements and functional transportation

**CM: What do you aim to do as a next generation fashion designers?**
SL: Personally, I would like to find a way to be a designer, but maybe a bit more lik an artist where you have multiple projects... projects for yourself, and creat collections, but also to collaborate. I also think this time of being in pandemic ha really emphasised those thoughts of how we can stop excessive growth. I think a young designers, we now have an opportunity to do things in our way that isn't a mass-produced. We don't have to follow in the system that existed before us. W can take an alternative way. It gave me confidence to continue to believe in mysel do what I do and I can just stay myself and be productive and creative, just shar my passion and love of what I do. I am very excited to try different things and lear from different things and discover what I like and what I don't like.

AR: My aim would be to create from a place of love first of all. Alot more peopl started from a place of compassion when they start creating. I think that woul give ring-effects in all parts of the industry that, in terms of how we treat eachothe in terms of how we produce.

AW: Especially for us young designers right now, it's very important to express ou own point of view. As a next generation, that's definitely something really importar because with global brands, sometimes the message can get slightly lost in thi cycle of consumerism. What young designers can bring specifically, the more tha they pursue their own vision, the more pure and the more controlled that they hav over something in which has more integrity. So I think I am really in support c people or young creative people pursuing their own individual practice as a thing What that can bring to a creative industry is invaluable.

DZ: I would like to continue my material research. I am releasing collections always pursuing resolution to biodegradable and tactics for fabrics makin

WHAT DO YOU WANT A MEANING FOR?
LIFE IS A DESIRE, NOT MEANING.

*photos_Courtesy of BALENCIAGA*

Suresnes-De Gaulle
Bac – Saint-Placide
70

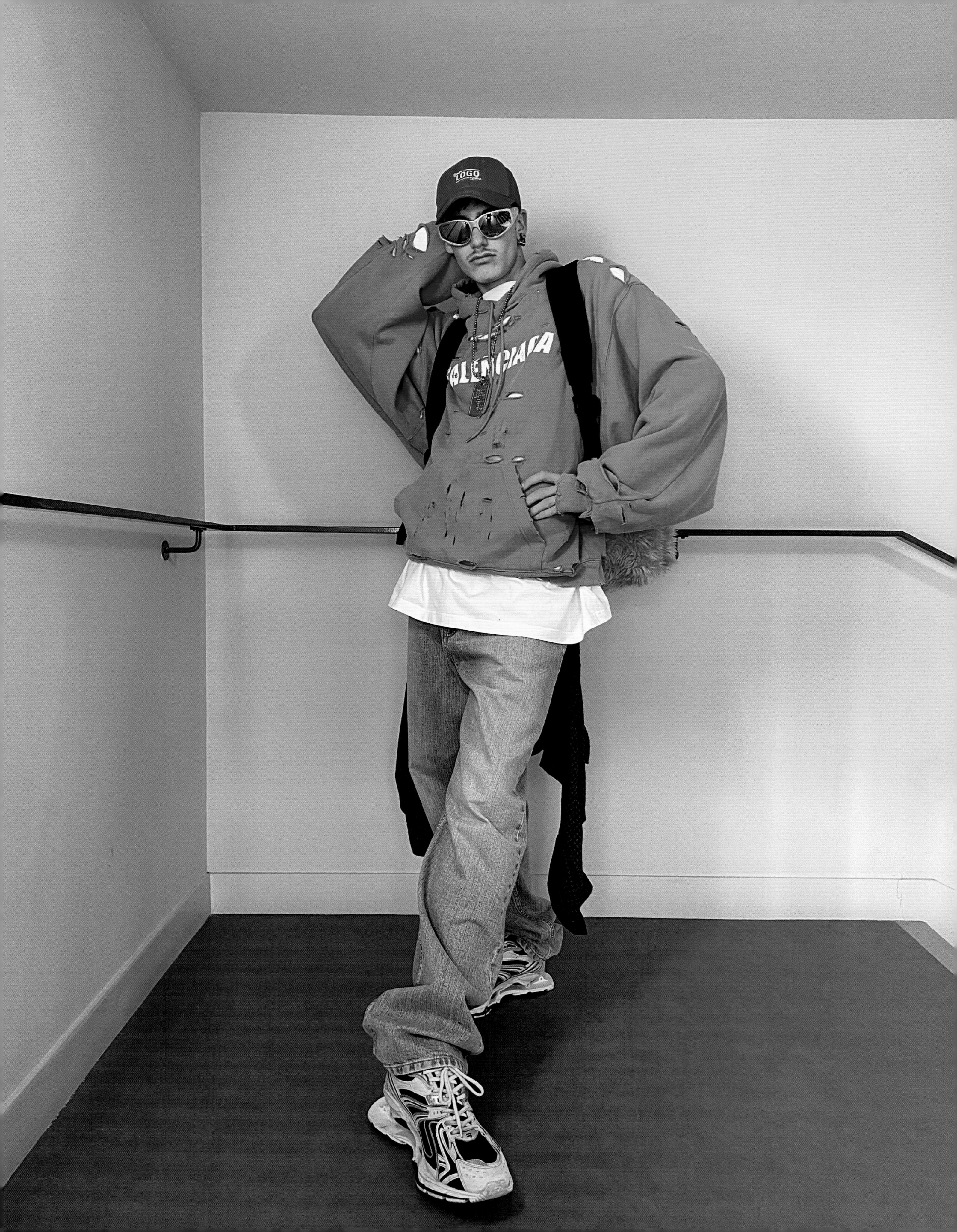
LOGO

NOT KNOWLEDGE

*all items by DIOR* ©Brett Lloyd for Dior

jacket, top &

op & trousers

# COMMON SENSE IS THE COLLECTION OF PREJUDICES ACQUIRED BY AGE 18

photos_Oscar Chang Anderson, Francis Kokoroko, Chris Cunningham & Jackie Nickerson

all items by **DIOR**

究極のエレガンスと新時代のマスキュリニティが投影された
ORのSUMMER 2021メンズ コレクション。叙情的な奥行きを
ップな明るさとエレガンスでコーティングした一連のワード
ーブは、ガーナのアクラ出身で、オーストリアのウィーンを拠点に活
するアーティスト、アモアコ ボアフォとのコラボレーションによる
のだ。起点となったのは、2019年、マイアミのルベル ミュージアム
おけるキム ジョーンズとアモアコの運命的な出逢いだという。

DIOR's Summer 2021 mens collection is a projection
ultimate elegance and new-age masculinity. The series
wardrobes filled with lyrical depth and a light pop and elegan
are the result of the collaboration with Amoako Boafo, an art
from Accra, Ghana, based in Vienna. The collaboration start
as a fateful encounter between Kim Jones and Boafo at t
Rubell Museum in Miami in 2019. Kim spent his childhood
African countries and has always been fascinated by Afric

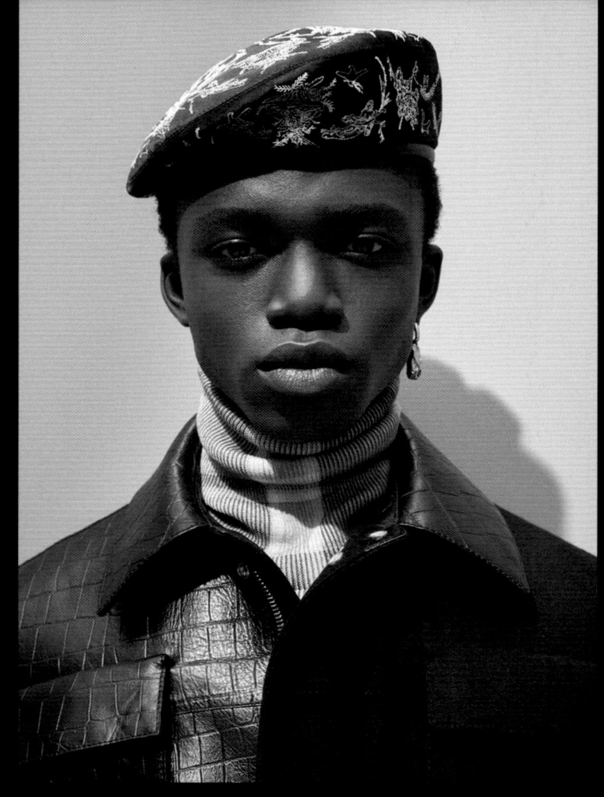

幼少期をアフリカ諸国で過ごし、アフリカのアートに魅せられてきたキムと、ファッションを重要なインスピレーションとするアモアコは、お互いに"一目惚れ"。そこから2人の文化的な対話がはじまり、今回の協業に結実した。黒人としての自身のアイデンティティを問い続けてきたアモアコは、黒人をモデルとする肖像画を得意とするが、その力強さと洗練性が同居するモダンな作品から溢れ出る強烈なメッセージが、稀代のストーリーテラーであるキムの心のクリ

art, and Amoako's work is inspired is fashion, and the match made in heaven lead to a cultural dialogue between the two. Boafo's work consists of portraits of black people as he contemplates his identity as a black man. The powerful and sophisticated strong message in his contemporary works stroke Kim's creative and storyteller heart. His strong brushstrokes applied by dipping his fingers directly in pigment and onto the canvas have become beautiful moiré patterned

BOY IN A BLACK & WHITE STRIPED SHIRT 2018
Oil on paper 100 x 80cm Courtesy Robert Projects

エイティブな琴線に触れたのだろう。指に直接顔料をつけてキャンバスに乗せることで生まれる力強い筆致は、美しいモアレ模様に。人物の肌を彩るヴィヴィッドでシュルレアルなネオンカラーをアクセントに採用し、絵筆の繊細なストロークはジャカードで表現。作品が放つ多角的な魅力は、キム独自の美意識とDIORのサヴォワールフェールにより軽快で着心地よいガーメントに翻訳され、壮大な

garments, and the vivid and surreal neon accents in the skin painted in delicate brush strokes are expressed in jacquards. The multifaceted appeal of Boafo's works are translated into a light and comfortable collection under Kim's unique sense of style and DIOR's savoir-faire, sublimating into a magnificent a pageant of modern menswear glory.

# LOUIS VUITTON FASHION EYE KYOTO by MAYUMI HOSOKURA

DO YOU WANT TO SEE MORE ?

LOUIS VUITTONと書籍。
その歴史は大の愛書家であり、書籍の収集家で知られる創業者の孫、ガストン-ルイ ヴィトンまで遡る。ガストン-ルイ ヴィトンは初めて出版業に参入し、新進気鋭のアーティストとのコラボレーションによる希少な版などを刊行。書き手、印刷技術、そしてアーティストをつないだと言われており、そのスピリッツはルイ ヴィトン パブリッシングとして、現代まで受け継がれている。今まで100タイトル以上を刊行しているLOUIS VUITTONの書籍のなかでも、旅をテーマにした3つのコレクション、『シティ ガイド』、『トラベルブック』、『ファッション アイ』は、コレクターも多い人気のシリーズ。特に『ファッション アイ』は、ファッション フォトグラファーの視点を通して捉えた街や地域、国の魅力が詰まったユニークなトラベル写真集。アップカミングな若手からファッション フォト界のレジェンドまで毎回異なるフォトグラファーを迎え、アイコニックな作品から未公開のアーカイブ作品までを1冊に収録。2016年のローンチ以来、現在23タイトルのラインナップとなっている。2021年3月には新たに『ファッション アイ ノルマンディ by ジャン モラル』及び、今回紹介する『ファッション アイ 京都 by 細倉 真弓』がシリーズに加わる。1979年京都生まれの細倉真弓の作品は、美しく有機的であり、官能的、そしてどこか夢のようでもある。そんな彼女の捉える"京都"は、カメレオンのように変化する捉えどころのない"ブルー"が印象的。ありふれた風景、自然、古式ゆかしい儀式、透明感あるティーンエイジャー… 彼女がカメラに収める被写体は知らぬ間にブルーに満たされる。美しい青色の表紙をめくり、神秘的で、どこか謎めいた儚さを感じさせる京都へと想いを馳せて…。
また、現在、原宿で開催されている『LOUIS VUITTON &』展内に設置される厳選されたLOUIS VUITTON製品を取り扱う特別なストアでは『ファッション アイ 京都』含め、LOUIS VUITTONのパブリケーションの世界も堪能できる。

The connection between LOUIS VUITTON and books dates back to the life of Louis Vuitton's grandson, Gaston-Louis Vuitton, who was known as a great bibliophile and book collector. Gaston-Louis Vuitton was the first to enter the publishing business in the family and published rare editions in collaboration with up-and-coming artists. He is said to have connected writers, printing techniques, and artists, and his spirit lives on as Louis Vuitton Publishing today. Among more than 100 titles LOUIS VUITTON has published up to this point, the three travel-themed collections, "City Guide", "Travel Book", and "Fashion Eye" remain the most popular series among amid collectors. "Fashion Eye" is an especially unique series with travel photography filled with the charm of each city, region, and country, captured through the perspective of fashion photographers. Each issue features different photographers, from up-and-coming young talents to fashion photography legends, and includes iconic photographs and previously unpublished archival works. Since its launch in 2016, the collection has grown to 23 titles. In March 2021, "Fashion Eye Normandie by Jean Moral" and "Fashion Eye Kyoto by Mayumi Hosokura" will be added to the collection. Mayumi Hosokura, born in Kyoto in 1979, is known for her beautiful, organic, sensual, and dreamlike works. The version of Kyoto she captures carries an elusive shade of blue that changes like a chameleon. From mundane day-to-day scenes, nature, ancient rituals to innocent teenagers, the subjects she captures are surrounded by layers of blue. Flip through the pages of the book with the beautiful blue cover to discover the mystical, mysterious and transient city of Kyoto. At the current "LOUIS VUITTON &" held in Harajuku, you will find the "Fashion Eye in Kyoto" at the special store which carries selected LOUIS VUITTON products and indulge in the world of LOUIS VUITTON publications.

『ファッション アイ 京都 by 細倉 真弓』
サイズ: 23.5 x 30.5 x 1.5 cm
ページ数: 96ページ
2ヶ国語版併記(仏語+英語)

"Fashion Eye in Kyoto by Mayumi Hosokura"
Size: 23.5 x 30.5 x 1.5 cm
No. of Pages: 96 pages
Bilingual (French & English)

A Stone in his Hand

View from Yasaka Shrine

Pink, Higashiyama-ku

FENDI
ROMA

FENDI
ROMA

FENDI
FENDI
ROMA

FENDI
ROMA
FENDI

FENDI
ROMA
FENDI
ROMA

# FENDI PEEKABOO ISEEU

photo_Daniele La Malfa

DO YOU WANT TO SEE MORE ?

## #FendiSS21

2009年春夏コレクションで華々しくデビューしたFENDIの"PEEKABOO"。以来、シーズンごとに素材や色、サイズなど新たなバリエーションを提案し、ファンを虜にしてきた。2014-2015年秋冬からはメンズラインも登場。これまでに発表したナイロン素材のモデルは、新時代のラグジュアリーとして話題を呼んだ。今回、2021年春夏メンズコレクションで満を持して発表された最新作"PEEKABOO ISEEU"は、ライニングの素材が最高級カーフレザー"クオイオ ローマ"になったり、内側のポケットが取り外し可能でリバーシブル仕様になっていたり、留め具がメタルフックからツイストロックに変更されていたり……。エレガンスをキープしながら、一層コンテンポラリーに、かつ機能的に進化を遂げている。

Each season since the spectacular debut of FENDI's "PEEKABOO" in the Spring/Summer 2009 collection, FENDI has been offering the iconic bag in various materials, colours, and has continued to captivate fans all over the world. From Fall/Winter 2014-2015 on, FENDI has introduced a menswear line, and the nylon model released until now has become widely known as starting a new era of luxury. The latest "PEEKABOO ISEEU", which was unveiled at the Spring/Summer 2021 menswear collection, features a quality calf leather "Cuoio Roma" lining, a removable and reversible inner pocket, and a new clasp. The clasp has been updated from a metal hook to a twist lock. The bag has evolved to become more contemporary and functional while maintaining its elegance.

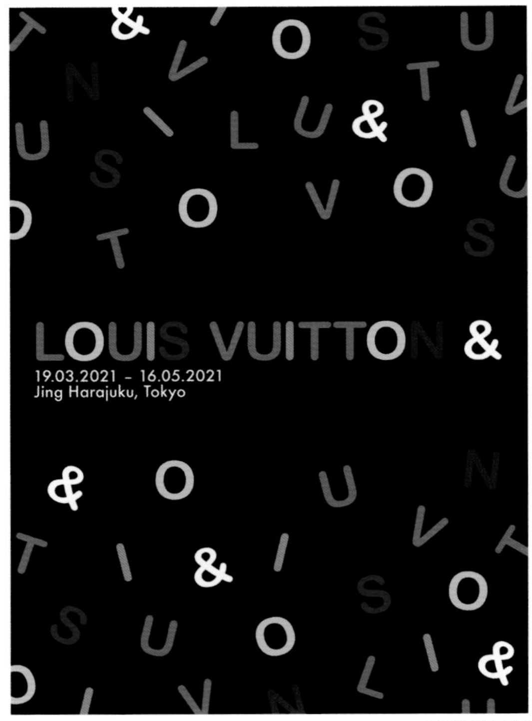




原宿 JINGでは、LOUIS VUITTONが重ねてきたアーティスティックなコラボレーションを中心に、メゾンの160余年の歴史を辿る展覧会『LOUIS VUITTON &』が5月16日まで開催中。川久保 玲、山本寛斎、藤原ヒロシ、草間彌生、NIGO®をはじめとする、日本の数多くのアーティストや、国際的に活躍するパーソナリティたちを讃え、LOUIS VUITTONとのコラボレーションを一堂に紹介。公式サイトより要予約。LOUIS VUITTON CLIENT SERVICE ☎ 0120 00 1854

DIORのスニーカーシリーズに新作"B28"が加わった。キム ジョーンズがFALL 2021コレクションのためにデザインしたスニーカーは、厚手のラバーソールが特徴的。メゾンを象徴するタイムレスなコード"ディオール オブリーク"モチーフで、写真のハイカット(¥135,000 ※予定価格)の他に、ローカット(¥120,000 ※予定価格)もあり、それぞれブラック&ベージュ、ブラック、ホワイトの3色展開。CHRISTIAN DIOR ☎ 0120 02 1947

CONVERSE ADDICTから、"JACK PURCELL CANVAS"(¥16,000)のブラックとカーキ、"ONE STAR SANDAL"(¥15,000)のサンドカラーといった新色が4月に発売予定。どちらもアーカイブのデザインを継承しつつ、VIBRAM® MRGAGRIPのアウトソールと、クッション性の優れたE.V.A.を採用し機能的にも昇華させたシューズとなっている。
CONVERSE INFORMATION CENTER ☎ 0120 819 217

MARNI初のポップアップストアが渋谷PARCO 1Fで、4月1日〜22日の期間限定で開催。"MARNIFESTO"と掲げられた2021年春夏コレクションから、ハンドペイントされたトランクバッグや日本限定のトートバッグ、Tシャツ、ウォレットなどが週替わりに展開。ウィメンズ、メンズともにラインナップ。ぜひ足を運んでみては?
MARNI JAPAN tel. 03 6416 1021

創設100周年を迎えるWELEDAから、新しくオーガニックヘアケアシリーズとして、ヘアワックス(30g ¥2,200)、ヘアミスト(115ml ¥2,400)、ヘアフォーム(100ml ¥2,400)の3種類が誕生。天然由来成分100%のトリートメント効果で、ツヤと潤いを与え、美しく健やかな髪に。
WELEDA JAPAN
☎ 0120 070 601

MASTER-PIECEからエコバッグ"STOREPACK"(¥4,000)が発売中。素材には環境負担低減した帝人フロンティア 株式会社のSTOLEX®を使用。メッシュのポケットの中にあるコードを引っ張ると素早く収納でき、付きで持ち運びも便利。黒、ベージュ、赤、黄、青、緑の6色展開。

MSPC
tel. 06 6265 2677

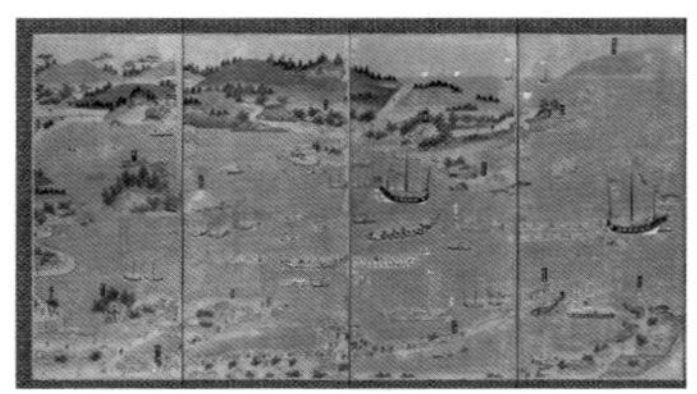

首里那覇港図屏風(部分) 19世紀 沖縄県立博物館・美術館蔵

国立歴史民俗博物館では、特集展示『海の帝国琉球—八重山・宮古・奄美からみた中世—』が5月9日まで開催。青磁、白磁、国宝の文書や重要文化財の梵鐘、屏風、絵図など400点以上の資料から、琉球帝国による侵攻と周辺島々の社会の変化、大航海時代より以前の知られざる八重山、宮古、奄美の世界など新たな歴史を見ることができる。
tel. 050 5541 8600
(ハローダイヤル)

福沢一郎《人》1936年 油彩・カンヴァス 東京国立近代美術館所蔵

諸橋近代美術館では、『ショック・オブ・ダリー サルバトール・ダリと日本の前衛—』が4月24日〜6月27日まで開催予定。見る者に強烈な印象を与えるダリの絵画。ダリの衝撃を最初に受け、それらを自らの創作に生かした日本の画家たちの表現を検証。日本における最初期のダリの影響の様相、日本のシュルレアリスム受容の一側面が明らかに。
MOROHASHI MUSEUM OF MODERN tel. 0241 37 1088

レオナール・フジタ(藤田嗣治)《タピスリーの裸婦》1923年 油彩/カンヴァス 126.0×96.0cm 京都国立近代美術館 ©Fondation Foujita / ADAGP, Paris & JASPAR, Tokyo, 2020 B0513

ポーラ美術館では、『フジター色彩への旅』が4月17日〜9月5日まで開催予定。レオナール フジタは旅先で目にした風景や人物、異国の歴史や風俗などに創作のインスピレーションを求め、パリ、南米、中米、北米、中国大陸や東南アジア、日本各地など、世界中を移動。フジタの旅による色彩の変遷、その生涯と画業の展開を見ることができる。
POLA MUSEUM OF ART
tel. 0460 84 2111

# MANY THINGS

MIKIMOTO COMME DES GARÇONSのコラボレーション ネックレス 第2弾が発売中。新作では反骨精神を表現するシルバーのセーフティー ピン、スタッズ、ファングと伝統的なMIKIMOTOパールのコンビネーション。相反するイメージが融合し、ひとつのデザインに。全7種類が発売中。
¥660,000
MIKIMOTO COMME DES GARÇONS
0120 868 254

雲の上の図書館 / YURURIゆすはら 2018
©Kawasumi・Kobayashi Kenji Photograph Office

高輪ゲートウェイ駅 2020 ©東日本旅客鉄道株式会社

東京国立近代美術館では、『隈研吾展 新しい公共性をつくるためのネコの5原則』が6月18日〜9月26日まで開催予定。各国に点在する隈作品の中から公共性の高い68件の建築を模型や写真、モックアップなどで紹介。その他、映像作品や前庭に展示されるトレーラーハウスなども。また、ネコの視点で都市の生活を見直すリサーチプロジェクト『東京計画2020 ネコちゃん建築の5656原則』も発表。
tel. 050 5541 8600（ハローダイヤル）

きょうも、これまでも
1987年から動物実験なしの配合

2020年、AĒSOPのサスティナビリティが権威ある3団体に認められた。1. AESOP本社のあるオーストラリアで2018年1月以降、CO2の排出量実質ゼロを達成し "Climate Active"認証を取得。2. リーピングバニー認証にて、製品に動物実験が行われていないことが認められる。3. 10月には"B corp認証"も取得。
AĒSOP JAPAN
tel. 03 6271 5605

BALENCIAGA SUMMER21で登場した新バッグライン"FLUFFY"。フェイクファー素材で、フロントのジップポケット部分にBALENCIAGAのロゴが施されたオーバーサイズのバックパック。
FLUFFY XXL BACKPACK（43 x 55 x 25cm）
¥128,000（※参考価格）
BALENCIAGA
CLIENT SERVICE
tel. 0120 992 136

1979年に登場したラグジュアリースポーツウォッチ"PIAGET POLO"から、スケルトンウォッチが登場。"PIAGET POLO"の骨太な外見を保ちつつ、デザイナーと技術者の技でこのモデル史上もっとも薄いケースを実現。PIAGETブルーのCVD加工されたムーブメントと、ガルバニック加工によるスレートグレー仕上げのムーブメントの2スタイル。
¥3,060,000
PIAGET 0120 73 1874

エシカル＆サスティナブルコスメブランドBOTANICANONに入浴剤が登場。鹿児島県大隅半島にある南大熊町の契約農家で無農薬栽培されたホーリーバジルを使用し、バイオ発酵エタノールで抽出した濃縮エキス100%。アーユルヴェーダで最高位のハーブとされるホーリーバジルの香りで、癒しのバスタイムを。
ホーリーバジル バスエッセンス（400ml）¥2,500
BOTANICANON
tel. 0994 24 3008

GREEN BEAVERから、天然由来成分100%、ヴィーガン処方の日焼け止め『ナチュラルサンスクリーン SPF40』が4月3日新発売。ノンナノ ウォータープルーフ、紫外線吸収剤不使用。SPF40ありながら白浮きせず、ラズベリー種子油やヒマワリ種子油が肌の潤いを保ってくれる。
（90ml）¥3,800
K2 INTERNATIONAL
tel. 086 270 7570

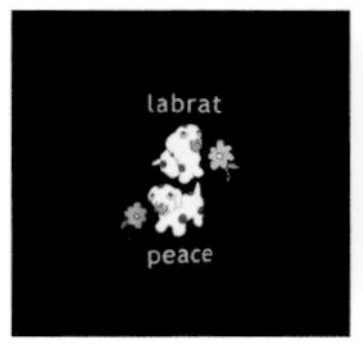

ニック アトキンスとLABRATがコラボレーション！ 1973年アメリカ ボストン生まれのニックは、SUPREMEのデザイナーであり、NYを拠点に活動するアーティスト。自身では『Electro Magnetic Studio』を主催。コラボレーションTシャツは4月発売予定。
LABRAT x Nick Atkins "peace" Tシャツ
¥6,000
LABRAT tel. 03 5474 6060

ジェンダーレスなインティメイトケアブランドMELLOW.が誕生。第1弾の商品『メロウドットインティメイトウォッシュ』は、デリケートゾーンが本来持つ常在菌の力と自浄作用を最大限に生かすための弱酸性pH値設計。また98.7%の植物由来成分で、15種の植物エキスと3種の精油を配合。乾燥から守り、潤いのある肌に仕上げてくれる。
（120ml）¥2,455
MELLOW.
tel. 050 7128 3366

LEDLENSERから新製品『Ledlenser ML6 Connect WL』が5月発売予定。人気のLEDランタン シリーズ"Ledlenser ML6"の後継機種で、Bluetoothでリモートコントロール可能、赤色点灯、ゆらぎモード、モバイルバッテリー機能などが搭載されバージョンアップ。屋内、屋外問わず様々な場面で活躍しそう。
¥11,800
LEDLENSER tel. 03 5637 7871

**PUBLISHER** KAORU SASAKI

**ISSUE DATE** 27th April 2021 BI-ANNUAL

**PUBLISHING** CUBE INC.
4F IL PALAZZINO OMOTESANDO 5-1-6 JINGUMAE SHIBUYA-KU TOKYO 150-0001 JAPAN
tel. 81 3 5468 1871 fax. 81 3 5468 1872 e-mail: info@commons-sense.net

**DISTRIBUTION** KAWADE SHOBO SHINSHA
2-32-2 SENDAGAYA SHIBUYA-KU TOKYO 151-0051 JAPAN
tel. 81 3 3404 1201 fax. 81 3 3404 6386 url: www.kawade.co.jp

**PRINTING** SOHOKKAI CO., LTD.
< TOKYO BRANCH >
2F KATO BUILDING 4-25-10 KOUTOUBASHI SUMIDA-KU TOKYO 130-0022 JAPAN
tel. 81 3 5625 7321 fax. 81 3 5625 7323
< HEAD FACTORY >
2-1 KOUGYOUDANCHI ASAHIKAWA HOKKAIDO 078-8272 JAPAN
tel. 81 166 36 5556 fax. 81 166 36 5657

 ISBN978-4-309-92224-9 C0070

**編集・発行人** 佐々木 香

**発行日** 2021年4月27日 年2回発行

**発行** CUBE INC.
〒150-0001 東京都渋谷区神宮前 5-1-6 イル パラッツィーノ表参道 4F
tel. 03 5468 1871 fax. 03 5468 1872 e-mail: info@commons-sense.net

**発売** 河出書房新社
〒151-0051 東京都渋谷区千駄ヶ谷 2-32-2
tel. 03 3404 1201 fax. 03 3404 6386 url: www.kawade.co.jp

**印刷** 株式会社 総北海 ＜東京支店＞
〒130-0022 東京都墨田区江東橋 4-25-10 加藤ビル 2F
tel. 03 5625 7321 fax. 03 5625 7323
＜本社工場＞〒078-8272 北海道旭川市工業団地2条1丁目
tel. 0166 36 5556 fax. 0166 36 5657

 ISBN978-4-309-92224-9 C0070